HARNESSING THE FOURTH INDUSTRIAL REVOLUTION THROUGH SKILLS DEVELOPMENT IN HIGH-GROWTH INDUSTRIES IN CENTRAL AND WEST ASIA

PAKISTAN

MAY 2023

ASIAN DEVELOPMENT BANK

ADB

Some rights reserved. Published in 2023.

ISBN 978-92-9270-134-5 (print); 978-92-9270-135-2 (electronic); 978-92-9270-136-9 (ebook)
Publication Stock No. TCS230155-3
DOI http://dx.doi.org/10.22617/TCS230155-3

The views expressed in this publication are those of the authors and do not necessarily reflect the views and policies of the Asian Development Bank (ADB) or its Board of Governors or the governments they represent.

ADB does not guarantee the accuracy of the data included in this publication and accepts no responsibility for any consequence of their use. The mention of specific companies or products of manufacturers does not imply that they are endorsed or recommended by ADB in preference to others of a similar nature that are not mentioned.

By making any designation of or reference to a particular territory or geographic area, or by using the term "country" in this publication, ADB does not intend to make any judgments as to the legal or other status of any territory or area.

Corrigenda to ADB publications may be found at http://www.adb.org/publications/corrigenda.

Notes:
In this publication, "$" refers to United States dollars.
ADB recognizes "Korea" and "South Korea" as the Republic of Korea and "Vietnam" as Viet Nam.

Cover design by Cleone Baradas.

Printed on recycled paper

Contents

Tables, Figures, and Boxes

Boxes

Foreword

The spectacular spread of Fourth Industrial Revolution (4IR) technologies globally has brought great upheavals and disruptions in labor markets. The *2020 Future of Jobs* report by the World Economic Forum estimated that by 2025, 85 million jobs may be displaced by a shift in the division of labor between humans and machines, while 97 million new roles could emerge that address the new realities of division of labor between humans, machines, and algorithms. The erstwhile fears of major job losses have given way to a more balanced discourse on amplifying the promise of technologies for sustainable development and human well-being and minimizing the peril of unemployment. While the new generation of disruptive technologies each have their own unique strength, it is the collective potential of these technologies to improve productivity and the quality of goods and services that have the greatest promise of influencing societal value and impact. Fusing the boundaries between the physical, digital, and biological worlds, the 4IR technologies that include artificial intelligence, robotics, the Internet of Things, 3D printing, genetic engineering, quantum computing, and machine learning, are fast becoming indispensable to modern work life, and indeed to the daily lives of citizens. The question is no longer how to prepare for 4IR technologies tomorrow but how to help individuals, firms, and societies today to effectively draw on them for greater productivity and prosperity.

To provide insights on the impact of 4IR on skills and jobs, the Asian Development Bank undertook the study "Harnessing the Fourth Industrial Revolution Through Skills Development in High-Growth Industries in Central and West Asia." The study suggests concrete pathways by which developing member countries can shape the transition of their economies to 4IR technologies to tap the potential for productivity and new jobs. It provides insights on opportunities, challenges, and promising approaches in 4IR for Azerbaijan, Pakistan, and Uzbekistan with specific focus on two industries in each country deemed important for growth, employment, and 4IR—transportation and storage and agro processing in Azerbaijan, information technology-business process outsourcing and textiles and garment manufacturing in Pakistan, and construction and textiles and garment manufacturing in Uzbekistan. The Central and West Asia region can benefit tremendously by tapping into 4IR technologies. It is important for the region to effectively manage the transition to 4IR technologies for greater economic diversification, moving up the global value chain and strengthening knowledge-based growth processes.

A key aspect of embracing 4IR technologies is to invest in appropriate skills. Based on a recent study by Amazon and Workplace Intelligence, 78% of Gen Z and millennial employees are concerned they lack the skills required to advance their career, and 58% are afraid that their skills have gone stale since the onset of the pandemic, and as many as 70% feel unprepared for the future of work. Hence the time to act on skills development is now, with ever-increasing demand for skills. The coronavirus disease (COVID-19) pandemic has caused incursions in business processes that have led to the acceleration of digital solutions in the marketplace. With the digital talent gap growing, there is a need for both public and private sector entities to invest in re-skilling and upskilling for new and transforming jobs due to adoption of technologies. The study stresses the importance of on-the-job training for 4IR technologies and the need for governments to embark on deliberate strategies for life-long-learning opportunities.

The study affirms a positive outlook to 4IR creating new opportunities for quality jobs. While many jobs will indeed be lost as a result of automation, new jobs will emerge through the adoption of technologies that will increase worker productivity and competitiveness of nations, thereby leading to greater prosperity. However, tapping such benefits is predicated on increasing investments in skills development and greater efforts by companies to upskill their workforce to perform new and higher order roles in complementarity with machines. The study has resulted in a suite of country reports for Azerbaijan, Pakistan, and Uzbekistan and a synthesis report that captures common elements across the three. The reports provide policy makers with evidence-based solutions for skills and talent development to strengthen the countries' readiness for a transition to 4IR.

The study highlights that while job losses will be real, a well-prepared 4IR strategy with industry transformation road maps that are recommended in the study can convert disruptions to opportunities to pivot the workforce to new and modern occupations. In light of post-COVID-19 realities, digital transformation and technology adoption can make enterprises more agile and responsive to changing market conditions.

We believe that 4IR technologies can not only bring greater economic value to enterprises and individuals, they can also help to strengthen the pathways for sustainable and inclusive development. There is more work to be done to explore and leverage the benefits at the intersection of digitalization and climate resilience and to scale up the deployment of 4IR technologies for equity and increasing opportunities for vulnerable populations. We welcome ideas and partnerships with stakeholders as we pursue this area of research toward concrete implementation and next level of analytical work.

Bruno Carrasco
Director General
Sustainable Development and
Climate Change Department
Asian Development Bank

Yevgeniy Zhukov
Director General
Central and West Asia
Department
Asian Development Bank

Preface and Acknowledgments

The Asian Development Bank (ADB) study "Harnessing the Fourth Industrial Revolution Through Skills Development in High-Growth Industries in Central and West Asia" addresses a crucial topic of great relevance to labor markets and jobs. At the heart of this study is the quest to better understand how disruptive technologies are influencing the nature of jobs and skills. Technologies of the Fourth Industrial Revolution (4IR) are influencing every sector and sphere of economies and societies, whether manufacturing or services. At the same time, business processes such as marketing, storage, transport, logistics, and payment mechanisms are greatly transformed with digital technologies. Business practices have been disrupted and reengineered through frontier technologies such as artificial intelligence, digital twins, robotics, and 3D printing.

We bring this piece of research to inform policy makers and practitioners of the implications of 4IR for future job markets. The study drew on various sources of secondary and primary data. It included surveys of employers and training institutions to assess their readiness for 4IR. The study presents analysis of data from online job portals from each of the countries covered in the study—Azerbaijan, Pakistan, and Uzbekistan—to assess trends in skills demand.

The study used a modeling exercise to estimate both job displacement and gains in select industries in the 3 countries. A review of the policy landscape based on benchmarks from international experiences provides the basis for the action points that developing countries can use to harness the potential of Industry 4.0 to increase productivity, facilitate skills development, and incentivize industry. The findings and recommendations from the study underscore the need for renewing skills development strategies with a full life cycle approach. This means that there are no degrees or certificates for life and regular upskilling is essential. The preponderant focus on institution-based training needs to give way to more flexible and multimodal training to include bootcamps, e-learning, and work-place based training. Training for digital skills at basic, intermediate, and higher levels needs a significant ramp up as workplaces undergo digital transformation. The benefits of 4IR can only be effectively harnessed if adequate investments are made in skills development.

The study was led by Shanti Jagannathan, in partnership with Eisuke Tajima and ADB team members. Rie Hiraoka and Brajesh Panth provided valuable guidance to the study. We thank the consultant team led by Fraser Thompson, director, AlphaBeta, for an excellent partnership in this study, together with Wan Ling Koh and Shivin Kohli. AlphaBeta's team developed the analytical model for the study and collaborated closely with ADB's team to bring new insights and directions and we are grateful for this professional collaboration. We thank Xin Long, Aziz Haydarov and Kevin Corbin from ADB headquarters and representatives of ADB resident missions in Azerbaijan, Pakistan, and Uzbekistan, respectively, for their valuable support and country-level consultations (Sabina Jafarova, Sanan Shabanov, Yuliya Hagverdiyeva and Elvin Imanov from Azerbaijan; Khuram Imtiaz and Rizwan Haider from Pakistan; and Farida Djumabaeva and Shahina Rismetova from Uzbekistan). Joehanne Kristal Santos and Evangelyn Medina from ADB provided timely coordination of meetings and activities during the study. Cherry Zafaralla copy edited this report. Dorothy Geronimo coordinated the editorial and publication process with ADB consultants: Maria Theresa Mercado (proofreading), Mariel Gabriel (proof checking), and Edith Creus (typesetting), and Cleone Baradas (cover design).

The study benefited greatly from enriching discussions with government representatives in the respective countries. Early workshops with government representatives and experts were held to inform the study process. The findings of the study were shared in country level workshops. Senior officials and key counterparts consulted are listed at the end of each country report. Tamerlan Tagiyev, Head, Center for Analysis and Coordination of the Fourth Industrial Revolution (Azerbaijan); Shabnum Sarfraz, member, Social Sector and Devolution, Planning Commission, Ministry of Planning, Development and Special Initiatives, Asadullah Faiz, member, Private Sector Development, Punjab Planning and Development Board, and Salman Shami, member, Private Sector Development, and Muhammad Haroon Naseer, additional director general, Punjab Skills Development Authority (Pakistan); and Oybek Shagazatov, head, Main Department of Cooperation with International Financial Organisations, Ministry of Investments and Foreign Trade (Uzbekistan). Several experts also contributed to the study—Amin Charkazov, Ramil Azmammadov (agro-processing) and Sabuhi Abdurahmanov (transport) from Azerbaijan; Allah Bakhsh Malik, Nasir Amin, Muhammad Asim Rehmat (information technology-business process outsourcing) and Muhammad Babar Ramzan (textiles) from Pakistan; and Shukhrathoja Amanov, Khabibullaev Shavkat Azamatovich (construction), and Umida Vakhidova (textile) from Uzbekistan.

We look forward to discussions in taking forward the study's policy recommendations.

Sungsup Ra

Sungsup Ra
Chief Sector Officer
Sustainable Development and
Climate Change Department
Asian Development Bank

Abbreviations

4IR	Fourth Industrial Revolution (or Industry 4.0)
ADB	Asian Development Bank
AR/VR	augmented reality and/or virtual reality
BAU	business-as-usual
COVID-19	coronavirus disease
ICT	information and communication technology
ILO	International Labour Organization
IT–BPO	information technology–business process outsourcing
IOT	Internet of Things
ITM	industry transformation map
PSDF	Punjab Skills Development Fund
PVTC	Punjab Vocational Training Council
SMEs	small and medium-sized enterprises
STEM	science, technology, engineering, and mathematics
TVET	technical and vocational education and training
TEVTA	Technical Education and Vocational Training Authority

Executive Summary

Labor markets globally have undergone tremendous disruptions and transformations from the adoption of fourth industrial revolution (4IR) technologies and digitalization. These trends have had positive and negative bearings on jobs—on one hand, enhancing productivity and improved business practices, and on the other hand, causing job losses and skills shortages. However, it is clear that these new technologies are reshaping labor markets, employment, and growth in significant ways. Nearly all occupations are impacted by the deployment of 4IR technologies, calling for timely policy actions to mitigate labor market disruptions and facilitate skills development and inclusion to help value addition to firm and businesses. It is anticipated that there will be both job losses and job gains from the deployment of 4IR technologies. To further boost job gains and help workers to remain employed in jobs that may be transformed by 4IR technologies, concrete efforts are needed to improve the readiness of industries and individuals to make the transition. Developing countries in particular need to ensure appropriate policy frameworks to prepare the workforce of tomorrow for 4IR to maintain and strengthen their comparative advantage in global markets.

Rapid advances in artificial intelligence (AI) and automation technologies have the potential to significantly disrupt labor markets. While AI and automation can augment the productivity of some workers, they can replace the work done by others and will likely transform almost all occupations at least to some degree. Rising automation is happening in a period of growing economic inequality, raising fears of mass technological unemployment and a renewed call for policy efforts to address the consequences of technological change.

The coronavirus disease (COVID-19) pandemic is further accelerating the digital transformation of businesses and jobs across all industries. It is anticipated that many new digital strategies adopted during COVID-19 may not completely disappear even beyond the pandemic. To better understand the implications of 4IR on the future of jobs and to assess the readiness of education and training institutions to prepare workers for future jobs, the Asian Development Bank (ADB) undertook the study *Assessing Implications of Industry 4.0 on Jobs and Skills in High-Growth Industries in Central Asia*, which seeks to capture the anticipated transformations in jobs, tasks, and skills, and to outline policy directions to prepare the workforce for future jobs, particularly in the post-COVID-19 world.

Scope and Methodology

The study comprises four volumes or reports covering three countries in the Central and West Asia region: Azerbaijan, Pakistan (with focus on Punjab), and Uzbekistan—including a synthesis report that draws together the findings from the three country studies. This report on Pakistan is the second of the four volumes, while the report on Azerbaijan is volume 1 and volume 3 is the report on Uzbekistan. The synthesis report outlines common areas of policy and action for Industry 4.0.

The study has the following features:

(i)   Two focus industries were selected in each country that are crucial for growth, employment, and 4IR, namely, agro-processing and transportation and storage in Azerbaijan; textile and garment manufacturing and information technology–business process outsourcing (IT–BPO) in Pakistan; and textile and garment manufacturing and construction in Uzbekistan.

For the focus industries, a survey of employers and training institutions, and analysis of data from online job portals to assess trends in skills demand and supply were conducted. The data collected was used to estimate job displacement and gains in the selected industries in each country through a logical model based on economic principles governing job creation and displacement.

(ii)   The policy landscape is also assessed. To understand gaps in the policy landscape in harnessing the potential of 4IR to increase productivity and create quality jobs, the study considered the comprehensiveness of policies in terms of stimulating 4IR adoption and worker reskilling efforts, creating new flexible qualification pathways, and building inclusiveness to extend the benefits of 4IR to all workers. The strength of implementation of policies, particularly against the backdrop of the COVID-19 pandemic, was also assessed.

Recommendations on how policy approaches toward 4IR could be strengthened are made. These focused particularly on the investments needed for skills and training, new approaches to deliver training, and other strategies and actions to enhance the readiness of each country's workforce for 4IR going forward. Surveys of employers and stakeholders were conducted between June and September 2021, and country-level data from 2020 (the latest for which full-year data is available) were used. To align all baselines for comparison, the survey data collected was assumed to be reflective of perspectives and circumstances as of end of 2020. The objective is to provide an illustrative view of how 4IR can impact jobs and skills in the three countries in the Central and West Asia region in the period between 2020 and 2025.

Key Findings for Punjab, Pakistan

This volume covers the key findings on effects of 4IR on jobs, tasks, and skills in the textile and garment manufacturing industry, and the IT–BPO industry in Punjab, Pakistan. Punjab is the second largest province by land area in Pakistan and a key driver of economic growth. In 2018, the province contributed approximately 54.2% of the national gross domestic product (GDP) and had a workforce of 40 million people, around 60% of the national labor force.

The focus industries were selected based on their importance to national employment and growth prospects, the degree of relevance of 4IR technologies to the industry, and alignment with national and regional growth plans. The textile and garment manufacturing industry is the largest employer after agriculture in Pakistan, providing employment to 7.8% of the population in 2018, and approximately 40% of the industrial labor force. Punjab is a key hub for the industry with approximately 70% of Pakistan's textile and garment manufacturers based in the province. Meanwhile, unlike the textile and garment manufacturing industry, the IT–BPO industry in Pakistan is at a relatively nascent stage of development, contributing to only 0.1% of national employment in 2018. However, there is significant potential for more workers to take up IT–BPO jobs. According to official statistics, Pakistan produces 20,000 information technology (IT) graduates and engineers annually, creating a strong human resource base for the industry. Nationwide, the IT–BPO industry recorded average employment growth of 6.5% between 2013 and 2018, significantly higher than the national growth rate of 2.7%. Both industries feature prominently in national and regional growth plans, including the Punjab Growth Strategy 2023.

The report finds that 4IR will have a transformational effect on jobs and skills in the textile and garment manufacturing and IT–BPO industries in Punjab with strong potential for positive gains in jobs and productivity, which can be reaped through adequate investments in skills and training.

Key findings of the study in Punjab include the following:

(i) Firms in both industries recognize the potential labor productivity gains from 4IR technologies but require additional support to adopt these technologies.

   (a) Textile and garment manufacturers estimate that the adoption of 4IR technologies could increase labor productivity (i.e., output per worker) by 35% between 2020 and 2025, while IT–BPO firms estimate a 45% increase.

   (b) However, only 35% of textile and garment manufacturers and 53% of IT–BPO firms surveyed have a strong understanding of 4IR technologies and their applications, and a significant proportion would require additional support to reap the gains from adopting such technologies.

(ii) The adoption of 4IR technologies can bring about overall job gains in the textile and garment manufacturing and IT–BPO industries in Punjab, but targeted policies are needed to ensure that job gains are equitably distributed.

   (a) The study estimates that the adoption of 4IR technologies in the textile and garment manufacturing and IT–BPO industries will create net job gains by comparing the displacement and productivity effects of adopting 4IR technologies. The number of new jobs created by productivity gains from adopting 4IR technologies will exceed the number of jobs displaced by automation.

   (b) As a result of 4IR technologies adoption, more jobs are expected to be created: 390,000 new jobs in Punjab's textile and garment manufacturing industry and 7,000 in the IT–BPO industry. This is over and beyond the business-as-usual job growth in a scenario without 4IR adoption, meaning that these jobs are fully attributable to 4IR adoption.

   (c) The new jobs created will be different from the jobs displaced, with employers in both industries assessing that the number of technical roles is likely to see the largest increase with 4IR adoption. In tandem with the change in types of jobs available, the skill needs of employers in both industries will change. Continual reskilling is critical to ensure that workers have access to quality jobs.

   (d) Another challenge specific to Punjab is the relatively low proportion of women in the textile and garment manufacturing workforce. Male workers account for close to two thirds of all garment workers in Pakistan and are expected to reap the bulk of the job gains from 4IR. Targeted policies are needed to ensure that job gains from 4IR are equitably distributed.

(iii) There is limited collaboration between employers and training institutions in the design of training curricula to ensure that workers are 4IR-ready.

   (a) The adoption of 4IR technologies will change the skill needs of employers in both industries by 2025. For textile and garment manufacturers, digital and information and communication technology (ICT) skills will be increasingly valued while creative thinking and complex-problem skills would become more important for IT–BPO employers. Training institutions need to engage employers regularly to ensure that training programs are aligned with changing skill needs.

   (b) There is limited collaboration in the design and implementation of training curricula and programs between employers and training institutions. Only 44% of training institutions gathered input from industry stakeholders to design curricula in 2020, and only 24% worked with employers to provide industry placements for staff for training purposes.

(iv)   There is limited focus among training institutions on 4IR-specific courses and use of 4IR technologies in knowledge delivery.

  (a)   Training institutions in Punjab demonstrated a high level of confidence in their preparedness for 4IR with a third of institutions surveyed strongly believing that they have a good understanding of the skills required. However, only 21% currently offer courses specific to 4IR technologies across all sectors. As 4IR technologies are increasingly adopted by industries, workers would likely require not just basic digital literacy but specific skills in areas such as AI or data analytics. Currently, only 21% of training institutions surveyed offer AI-related courses across all sectors. However, over 50% of firms in the textile and garment manufacturing and IT–BPO industries currently deploy such technologies.

  (b)   Industry 4.0 technologies can also help training institutions in Punjab improve their knowledge delivery, particularly against the backdrop of changing movement restrictions and school closures forced by the COVID-19 pandemic, when 75% of training institutions indicated that they had to close fully for some time due to inability to conduct in-person training. 4IR technologies can help to improve the effectiveness of training, particularly for courses delivered virtually. However, only 20% of training institutions currently use virtual reality or augmented reality tools to deliver courses.

(v)   The quality of training and job search support given to graduates vary across institutions.

  (a)   About 85% of employers surveyed agreed that there is significant variance in the quality of graduates from different training institutions and close to half of training institutions cited quality assurance mechanisms as a useful policy lever.

  (b)   The support given to graduates on their job search also varies between training institutions. Only 54% of training institutions surveyed provide information on job openings to their students although limited job opportunities and information asymmetry in the labor market were cited by training institutions as the most common reasons for their graduates being unable to find jobs.

(vi)   Punjab has adopted a range of strategies to prepare its workforce for 4IR, but there are some gaps in terms of implementation.

  (a)   A thorough scan of policies and programs across the government, industry, and civil society in Punjab reveals that while the Punjab Growth Strategy 2023 sets out the broad vision toward a knowledge-based economy, a 4IR-focused vision that provides a framework for the adoption of 4IR technologies across all industries has not been formulated. Industry-specific road maps that integrate 4IR trends and skills development needs and align stakeholders across government, industry, and training providers have also not been developed.

  (b)   In terms of policy coverage, there are various initiatives in Punjab aimed at upskilling the workforce digitally and incentivizing workers to participate in skills development. Some programs to build stronger alignment between industry and training providers and encourage a focus on skills over qualifications in labor markets are also in place. However, there is scope for existing policy coverage to be strengthened, particularly to build effective lifelong learning models. Thirty-nine percent of the working population in Punjab is illiterate and may not meet the educational qualification requirements of formal skills training providers to access reskilling opportunities (ADB 2020).

Key Recommendations and Way Forward

Drawing on the survey findings of employers and training institutions as well as the policy assessment, this report identifies seven recommendations for Punjab to strengthen its preparedness for 4IR, the specific actions of which include the following:

(i) Develop 4IR adoption road maps for key sectors.

   (a) While the Punjab Growth Strategy 2023 sets out the broad vision toward a knowledge-based economy, a 4IR-focused vision that provides a framework for the adoption of 4IR technologies across all industries has not been formulated. Industry-specific road maps that integrate 4IR trends and skills development needs and align stakeholders across government, industry, and training providers have also not been developed.

   (b) Punjab could consider the development of Industry Transformation Maps (ITMs) like in Singapore, which provide information on technology impacts, career pathways, skills required for different occupations, and reskilling options for different industries. The Punjab Planning and Development Board could coordinate such efforts across industries, and line agencies could be tasked with engaging industry stakeholders to understand their growth plans and challenges, as well as create industry-specific road maps, including for the textile and garment manufacturing and IT–BPO industries over an implementation time frame of 12–36 months.

(ii) Develop innovative job-matching initiatives and platforms.

   (a) The lack of job opportunities and labor market information asymmetries were cited by training institutions as key barriers to graduates finding jobs. Many employers in Pakistan continue to rely on informal channels or personal networks to employ workers. Policy makers in Punjab can consider innovative approaches to improve job-matching between employers and prospective workers.

   (b) The Punjab Department of Labour could consider working with technology agencies such as the Punjab Information Technology Board to launch platforms incorporating AI or Big Data technologies that could help job seekers or recruiters to easily sift through the available opportunities or candidates. This could potentially be implemented in less than 12 months.

(iii) Strengthen existing systems for recognition of prior learning.

   (a) In the textile and garment manufacturing industry, workers in manual job roles face the highest risk of displacement while jobs are most likely to be created in technical roles. In Punjab, 39% of the working population is illiterate and workers in manual or elementary jobs roles could face challenges in accessing reskilling opportunities, as they are unable to meet the educational qualification requirements of skills training providers.

   (b) Punjab could strengthen mechanisms for the recognition of prior learning and institute programs to promote continual skills upgrading. Policy makers could take the Malaysian Skills Certification Program as a reference, under which skill certificates are granted to workers without formal educational qualifications but who have obtained relevant knowledge, experience, and skills in the workplace to enhance their career prospects. This could be implemented by agencies such as the Punjab Skills Development Authority and Trade Testing Board of Punjab within a time frame of 12–36 months.

(iv) Promote adoption of 4IR technologies in knowledge delivery.

(a) Industry 4.0 technologies can also help training institutions in Punjab improve their knowledge delivery. This is particularly important against the backdrop of movement restrictions and school closures forced by the COVID-19 pandemic. While close to half of training institutions pivoted to offering courses online to continue training during the closures, only a small proportion of institutions use interactive videos, self-learning online modules, or virtual reality (VR) technologies in their training in 2020.

(b) Case studies in other countries show that 4IR technologies have substantial potential to enhance the delivery of virtual learning as well as classroom learning and can be adopted more widely by Punjab's training institutions, led by entities such as the Punjab Vocational Training Council and the Punjab Technical Education and Vocational Training Authority, and implemented within a time frame of 12–36 months. Collaboration with the private sector could help to address the resource constraints faced by training institutions in Punjab in the adoption of 4IR technologies and policy makers could devise programs or incentives to support such collaboration.

(v) Strengthen relevance of industry apprenticeships and internships.

(a) In Punjab, 73% of training institutions work with employers to organize workplace-based training for students and over half organize industry apprenticeships. Since 2020, the Higher Education Commission has made internships for university students mandatory, and all undergraduates are required to undergo a 9-week internship as part of course requirements. A new Apprenticeship Act that was recently passed further aims to strengthen the relevance of apprenticeships. Policies and frameworks to enhance the relevance of industry apprenticeship and internships, drawing on best practices in Singapore and the United Kingdom, could strengthen the effectiveness of existing internship and apprenticeship programs. In the United Kingdom, apprenticeship standards are developed by an employer group, the Institute for Apprenticeships, under the sponsorship of the Department of Education. The apprenticeship standard created in collaboration with employers sets out the skills, knowledge, and behaviors required of a qualified worker, and another document sets out how these are to be assessed at the end of the apprenticeship program.

(b) To set up similar apprenticeship standards and assessment methods, the Punjab Skills Development Authority could work with industry associations such as the All Pakistan Textile Mills Association for the textile and garment manufacturing industry, and the Pakistan Software Houses Association for IT, ITES, and IT–BPO. This could be implemented within a time frame of 12–36 months.

(vi) Adopt inclusive 4IR reskilling policies.

(a) Women make up approximately half of Pakistan's population but were less than a quarter of the workforce in 2018. In the ICT industry, women comprised less than 5% of the industry's workforce in 2018. This research shows that the adoption of 4IR technologies will exacerbate the inequalities in the IT–BPO sector. The number of new jobs expected to be gained by male workers due to 4IR adoption will exceed the number of jobs expected to be gained by female workers by 9.6 times by 2025.

(b) Policy makers in Punjab led by the Punjab Planning and Development Board can consider initiatives to build stronger digital literacy among women, and allow women to empower themselves through technology. India's Disha project that supports various programs to provide career counseling, skills training, and job placements for women, including the "Coding Mum" program in Indonesia are a useful reference for policy makers in Punjab. Such policies could be implemented within a time frame of 12–36 months.

(vii)   Ensure responsiveness of education systems to changing skills needs.

(a)   The study reveals that the skills sought by employers will change significantly with the adoption of 4IR technologies. Creative thinking or design skills, and complex problem-solving skills will become the most sought-after skills for IT–BPO employers by 2025, while digital and ICT skills will become most valued by employers in the textile and garment manufacturing industry. Soft skills such as complex problem solving and critical thinking will become more valued by employers than traditional skillsets such as written communication. Punjab could adopt best practices in the Republic of Korea to strengthen the responsiveness of primary and secondary curricula and teaching pedagogy to changing skill needs created by 4IR. In the Republic of Korea, the national curriculum framework is reviewed every 5–10 years to ensure that educational curricula reflect updated learning needs in tandem with the emerging demands of the labor market. The 2015 revision led to the introduction of both "soft" skills (such as creative thinking) as well as technical ICT skills into student learning outcomes.

(b)   In Punjab, the School Education Department can lead efforts to update education curricula and ensure that teachers are provided with sufficient training to deliver the curricula, in close coordination with the Punjab Planning and Development Board to understand future skill needs. Such efforts could be implemented within a time frame of 12–36 months.

While these recommendations apply to both the textile and garment manufacturing and IT–BPO industries, a set of priorities unique to each industry should be considered when implementing the respective recommendations.

Textile and Garment Manufacturing Industry

Companies in the textile and garment manufacturing industry have limited understanding of 4IR technologies and workers are likely to be lower-skilled and face more challenges in being retrained to take on technical roles. For instance, only 35% of textile and garment manufacturers have a good understanding of 4IR technologies and applications. The development of an industry transformation road map that addresses the limited awareness of 4IR technologies among firms and sets out measures to increase awareness and deployment must be a priority for policy makers. In addition, workers in manual roles are expected to face a higher risk of displacement while job gains are expected in technical roles, with over 70% of textile and garment employers expecting the proportion of technical roles to increase by 2025. This suggests that existing frameworks for recognition of prior learning would need to be strengthened. Job gains from 4IR will also largely benefit male workers due to low female workforce participation in the textile and garment manufacturing sector, as one-third of garment workers in Pakistan are female currently. Inclusive 4IR reskilling policies would need to be adopted to distribute the gains of 4IR more evenly.

Information Technology–Business Process Outsourcing Industry

The IT–BPO industry has a stronger understanding of 4IR technologies compared to the textile and garment manufacturing industry and skill needs are likely to change rapidly. Creative thinking and complex problem-solving skills are expected to become more important to employers in the IT–BPO industry by 2025 and school curricula would need to evolve rapidly to meet these new skill needs. At the same time, misalignment on skills demand between training institutions and employers would need to be addressed. The surveys suggest that 75% of IT-BPO employers plan to adopt AR/VR technologies but only 13% of training institutions offer courses in AR/VR technologies. It would therefore be critical for employers, training institutions, and other stakeholders to work closely to strengthen the relevance of industry apprenticeships and internships and address the immediate shortage of skilled talent trained in the latest technologies, as reported by local stakeholders such as the Pakistan Software Houses Association for IT and ITeS (P@SHA). In addition, the participation of women in careers in science, technology, engineering, and mathematics (STEM) would also need to be increased through inclusive 4IR reskilling policies, as 4IR could exacerbate the challenges faced by female workers in the IT–BPO industry as the number of technical jobs increase.

1 The Industry 4.0 Skills Challenge

This chapter investigates the demand and supply of skills driven by the adoption of technologies that are hallmarks of the Fourth Industrial Revolution (4IR or Industry 4.0), such as Internet of Things (IOT), artificial intelligence (AI), additive manufacturing, robotics, and Big Data. The analysis uses a range of data from employer surveys, expert interviews, online job board data, and national labor market statistics for the textile and garment manufacturing and information technology (IT)–business process outsourcing (BPO) industries in Punjab, Pakistan, and projects implications for 4IR adoption between 2020 to 2025.

The analysis shows that the adoption of 4IR technologies can bring about quality jobs for workers if strong policies to encourage firms to adopt 4IR and encourage worker reskilling are implemented. The adoption of 4IR could create 11% more jobs, over and beyond business-as-usual (BAU) growth rates in the textile and garment manufacturing industry; and 18% more jobs in the IT–BPO industry by 2025. For these gains to be realized, policies must be implemented to encourage firms to adopt 4IR technologies and build awareness of the digital tools available.

Strong reskilling policies will be needed to realize these job gains, and to support workers displaced by 4IR. To enable the adoption of 4IR technologies, employers expect that a larger proportion of the workforce in both industries will be concentrated in technical occupations by 2025. There will be significant changes in the relative importance of various skill sets for workers in each industry as the distribution of occupations changes. Digital and/or information and communication technology skills will be increasingly valued by firms in the textile and garment manufacturing industry while creative thinking and design skills will be increasingly sought after by IT–BPO employers. Strong skills development policies are needed to help workers reskill, particularly those in occupations prone to displacement. These policies are also needed to build a workforce able to support the adoption of 4IR technologies.

A. Industry 4.0 and Its Relevance for Punjab, Pakistan

The Fourth Industrial Revolution is poised to fundamentally change the future of work. 4IR can be described as the advent of "cyber-physical systems" involving entirely new capabilities for people and machines,[1] wherein new technologies, such as IOT, AI, additive manufacturing, robotics, and Big Data analysis among others, become embedded within societies. 4IR is fundamentally different from past industrial revolutions in its potential implications for economies and the workforce.

---

[1] World Economic Forum. What is the Fourth Industrial Revolution? https://www.weforum.org/agenda/2016/01/what-is-the-fourth-industrial-revolution/.

National and regional policy makers recognize the potential transformational impact of ICT and other emerging technologies. The Punjab government has adopted a series of policies and road maps that set out Punjab's vision to build a vibrant ICT sector and skilled workforce, which will enable its shift toward inclusive knowledge-based economic growth. The Punjab Growth Strategy 2023 (Planning and Development Board 2019) identifies the development of human capital as a key pillar for growth. The Punjab Information Technology Policy 2018 provides a road map for the province's development into a regional ICT hub through initiatives that cover the promotion of entrepreneurship, support for the ICT industry, and improvement of digital literacy for vulnerable groups (Punjab Information Technology Board 2018).

Understanding how the skills landscape is likely to change under 4IR is becoming more difficult in light of how rapidly technology is developing and being adopted. This is particularly so as these changes accelerated against the backdrop of the coronavirus disease (COVID-19) pandemic. This means traditional approaches of assessing skill gaps, often relying on time-intensive processes to collect data that quickly become outdated, may no longer be suitable.

This report explores a new approach to understanding the labor market implications of 4IR. Some of the key design aspects examined are as follows:

(i) **Use of primary and secondary local data.** This report utilizes a variety of local data sources, including data from the Pakistan Bureau of Statistics (PBS); surveys conducted with 52 employers in the textile and garment manufacturing, 51 employers from IT–BPO industries, and 71 employers from training institutions in Punjab, as well as interviews with local experts and key stakeholders. A summary of the primary data sources used is in Table 1.

Table 1: Primary Data Sources Used

Employer surveys	52 textile and garment manufacturers and 51 IT–BPO firms in Punjab, Pakistan were surveyed. The surveys were carried out by Ipsos (Global market research company).
Training institution surveys	71 training institutions in Punjab, Pakistan were surveyed. These include public and private institutions of higher learning as well as technical and vocational education and training institutes. Of the training institutions surveyed, more than 90% trained at least 100 students per year. The surveys were carried out by Ipsos (Global market research company).
Online job portal analysis	The analysis covered 86 job listings in the textile and garment manufacturing industry and 1,237 job listings in the IT–BPO industry obtained from the job portal Rozee.pk in June 2021.

4IR = Fourth Industrial Revolution, IT–BPO = information technology–business process outsourcing.
Source: Asian Development Bank Sustainable Development and Climate Change Department.

(ii) **Use of current market information.** Given the rapid changes in the labor market, labor market surveys can become quickly obsolete. To provide an updated snapshot of skills needs, this study uses information on skill profiles for current jobs advertised in major online job portals.[2] Machine-learning algorithms were applied to analyze data from local job portals to understand the skills demanded by employers in the two industries.

(iii) **Focus on both demand and supply.** The study considers changes in the demand of skills brought about by the adoption of 4IR technologies, as well as the supply of skills, and the readiness of the training landscape to upskill and reskill workers for 4IR.

---

[2] The analysis covered 86 job listings in the textile and garment manufacturing industry and 1,237 job listings in the IT–BPO industry from the job portal Rozee.pk (accessed June 2021).

B. Industry Selection

Two industries were selected in each country to conduct further analysis of the implications of 4IR adoption for the demand and supply of skills. A two-step methodology was used to select the industries:

(i)   **Shortlisting industries prioritized by the federal government or Punjab government for future growth or for 4IR application.** This included reviewing the Punjab Growth Strategy 2023, Digital Pakistan Policy, and Pakistan Vision 2025, among other policy documents.

(ii)   Scoring and ranking shortlisted industries according to a set of criteria, including the following:

   (a)   How significant is the industry's contribution to the country's employment?
   (b)   Does it exhibit strong recent employment growth?
   (c)   Are its exports internationally competitive?
   (d)   How relevant are 4IR technologies to the industry?
   (e)   Is the relevant data available for the industry analysis?

The shortlisted industries were then tested with stakeholders across government, industry, and civil society during consultation workshops and briefings conducted from April to June 2021. Based on this process, two industries were selected for the analysis in Punjab, Pakistan.

(i)   **Textile and garment manufacturing**. The textile and garment manufacturing industry is the largest employer after agriculture in Pakistan, contributing 7.8% to national employment in 2018 (Finance Division 2021). Between 2013 to 2018, the industry recorded average employment growth of 4.7%, significantly above the national growth rate of 2.7%.[3] The textile and garment manufacturing value chain in Pakistan is comprehensive, and includes various stages of processing from cotton production to spinning, fabric manufacturing, dyeing, and sewing of garments. There is potential for 4IR technologies such as IOT, AI, additive manufacturing, autonomous robots, and Big Data analytics to provide value across the different stages. Official statistics estimate that barring seasonal and cyclical fluctuations, textile products are around 60% of national exports, with cotton products taking up the largest share (Finance Division 2021). In particular, Punjab is a key hub for the textile and garment manufacturing industry with approximately 70% of textile and garment manufacturers in the country based in the province.[4] National and provincial policies recognize the importance of the industry to the future growth of Pakistan and Punjab. The national Textile and Apparel Policy, 2020–2025 aims to boost the production and exports of value-added textile products (Khan 2021). It aims to increase textile exports from $13 billion in 2018 to $25 billion in 2025 and reach $50 billion by 2030 (Mustafa 2020). The Punjab Growth Strategy 2023 sets out plans to modernize textile and garment value chains to develop higher value-added products and enhance exports (Planning and Development Board 2019).

(ii)   **Information technology and business process outsourcing.** The IT–BPO industry in Pakistan is at a nascent stage of development. In 2018, the industry contributed only 0.1% to national employment and the entire ICT industry contributed only 0.53% to Punjab's employment. However, there is significant potential for more workers to take up jobs in the sector. According to official statistics, Pakistan produces 20,000 ICT graduates and engineers annually, creating a strong human resource base for the industry (*The News* 2019). Nationwide, the IT–BPO industry recorded average employment growth of 6.5%

---

[3]   Sources include Pakistan Bureau of Statistics, ILO statistics, WTT council, and AlphaBeta analysis.
[4]   Punjab Board of Investment and Trade. Textile Industries. http://www.pbit.gop.pk/textile_ind.

between 2013 and 2018, significantly higher than the national growth rate of 2.7%.[5] The IT–BPO industry has been identified as a key driver of future growth as Pakistan and Punjab progress toward a knowledge-based economy. Nationwide, IT exports increased 59% year-on-year to $200 million in May 2021 compared to $126 million in May 2020, with the growth attributed to a surge in digital freelancing due to the COVID-19 pandemic. The government is targeting to take IT exports to $10 billion by 2025, backed by policies such as the exemption of income tax on IT services exports till 2025.[6] In Punjab, the Punjab IT Policy 2018 and Punjab Growth Strategy 2023 set out plans to grow the ICT and IT services industry into one of the province's key generators of employment (Punjab Information Technology Board 2018).

C. Textile and Garment Manufacturing Industry

Industry 4.0 technologies pose significant potential for textile and garment manufacturers to increase their labor productivity. There are estimates that the full adoption of 4IR technologies across the textile and garment manufacturing value chain could reduce labor time by around 40% to 70% (McKinsey & Company 2018). Research in Viet Nam estimates that full automation could reduce the production time of a shirt by 18–33 times compared to manual sewing (Institute for Workers and Trade Unions 2020). Firms in Viet Nam are in the early stages of adopting automation at some stages of the production process but past research reveals that there is active discourse on 4IR within the industry that impacts the mindsets of managers who are considering the application of new technologies to improve product quality and productivity (Institute for Workers and Trade Unions 2020). Expectations among textile and garment manufacturers in Punjab are more conservative, likely due to nascent understanding of 4IR technologies, with only 35% of firms surveyed having a good understanding of 4IR technologies. Overall, firms expect 4IR technologies to increase output per worker by only 35% by 2025.

The analysis shows that if these productivity gains are realized, there will be a significant number of new jobs created for workers in Punjab by 2025. This report estimates that 390,000 net new jobs or the equivalent of 11% of the textile and garment manufacturing jobs in 2020 could be created between 2020 and 2025, if 4IR technologies are fully adopted by the industry. This is over and beyond the jobs created under a BAU growth scenario.

The 390,000 net new jobs created are a function of 1.52 million jobs created and 1.13 million jobs displaced with the adoption of 4IR technologies. New jobs created will be different from the jobs displaced and require different skill sets. Digital and/or ICT skills as well as adaptive learning capabilities will be increasingly sought by textile and garment manufacturing employers. Therefore, to ensure that Punjab's textile and garment manufacturing firms and workers can reap the benefits of the shift toward 4IR, it is critical that policy makers are focused on both encouraging the adoption of 4IR technologies among firms; and promoting continual reskilling among workers.

Relevance to Industry 4.0

There are various 4IR technologies relevant to the textile and garment manufacturing industry, ranging from robotics technology to additive manufacturing processes that enable the mass customization of products.

---

[5]   Sources: Pakistan Bureau of Statistics, ILO statistics, WTT council, AlphaBeta analysis.

[6]   Invest Pakistan. 2018. Income Tax Exemption on Exports of Computer Software or IT services or ITes. https://invest.gov.pk/node/1253; and Baloch.

Some key 4IR technologies and their applications in the textile and garment manufacturing industry include the following:

(i) **Internet of Things.** The IOT refers to networks of sensors and actuators embedded in machines and other physical objects that connect with one another and the internet. The textile manufacturing process has many stages such as spinning, weaving, dyeing, printing, finishing, and fabric manufacturing, all of which require close monitoring. An IOT-integrated system, enabled by sensors and drones, can help to provide real-time data across these stages and identify potential bottlenecks. One example is an IOT-enabled weaving unit that can synchronize every stage of the weaving process from yarn processing and inventory to production monitoring, right up to shipment of the finished fabric. It can manage the yarn inventory and optimize the production schedule, among other functions.[7] Egyptian startup Garment IO uses IOT technologies to transform clothing factories. Every worker is given a smart terminal and an electronic card. Once a batch of work has been completed, such as sewing an order of shirts, the workers scan their card on the terminal and the tag attached to the bundle of shirts. The smart terminal logs what order the worker completed, how long it took them, and how many more orders they have left to complete. This information is then uploaded to a cloud that managers can access in real-time along with detailed breakdowns of how the factory line is performing. This allows factories to identify and eliminate any potential bottlenecks quickly (Waya 2019).

(ii) **Artificial intelligence.** Artificial intelligence gives machines the ability to learn and act intelligently and carry out a wide range of human-like processes. This means they can make decisions, carry out tasks, and even predict future outcomes based on what they learn from the data. For instance, AI technology can be used to carry out fabric grading, and quality checks on fabrics and ensure that the colors of the finished textile match with the originally designed colors. Researchers in Hong Kong, China developed WiseEye, an AI-based automated textile inspection system to help manufacturers instantly detect defects in fabrics during production. WiseEye can reduce loss and wastage due to faulty textiles by 90% as compared to human inspection (Del Buono 2018). In Pakistan, firms such as Masood Textile Mills and Amami Clothing use CLO, a 3D garment simulation software, in their product development process. The software can be used to create virtual prototypes and modeling to replace physical prototypes and reduce the overall time and cost of the design process.[8] Retailers also use AI technologies to improve the online shopping experience. The AI-based startup Lalaland creates fully AI-generated fashion models that online shoppers can customize to look like themselves. By tailoring the "virtual fitting" experience, Lalaland's technology claims to be able to achieve a 15% increase in sales and a 10% reduction in returns for online retailers (Nicholls-Lee 2021).

(iii) **Additive manufacturing.** Additive manufacturing technologies or 3D printing produce physical objects from digital models by adding thin layers of material in succession. The process cuts down material waste and improves production efficiency. Additive manufacturing can be used for the creation of customized clothing, complex designs in garments and accessories, and prototypes. For example, sportswear brand Adidas is leveraging 3D printing technology and robotics to produce footwear modeled to the exact contours of an individual runner's foot (Hanaphy 2020). Apparel brand Ministry of Supply can produce a customized blazer in just 90 minutes using 3D printing technology, while reducing fabric waste in production by about 35% compared to traditional techniques (CB Insights 2022).

(iv) **Autonomous robots.** Autonomous robots are intelligent machines capable of performing tasks with a high degree of autonomy. There are several applications of robotics in this sector, including using robotic arms for repetitive processes such as weaving and sewing, needle positioning, fabric adjustment. Sewing

---

[7] Clarion Technology. *How IOT Transforms the Way to a More Sustainable Textile Manufacturing.* https://www.clariontech.com/blog/how-iot-transforms-the-way-to-a-more-sustainable-textile-manufacturing.

[8] Sources: Clo3d. Design Smarter. https://www.clo3d.com/; local expert.

is currently the most labor-intensive step in creating a garment, accounting for more than half the total labor time per garment. The potential for labor reduction varies by garment type, but as much as 90% of the sewing processes can potentially be automated (ACT/EMP and International Labour Organization [ILO] 2017). Studies show that full automation can reduce the production time of a shirt by 18–33 times compared to manual sewing (Institute for Workers and Trade Unions 2020). Autonomous robots also help to streamline other processes in the textile manufacturing production chain. For instance, Lahore-based garment manufacturer, Combined Fabrics, uses advanced cutting equipment from Tukatech, a leading provider of fashion software and machinery (Tukatech 2021). The automated precision cutting machines allow employees to be deployed to other tasks and reduces wastage by 12%–14% per garment.

(v)   **Big Data analytics.** Big Data analytics is the use of advanced analytic techniques on large, diverse data sets. In the textile and garment manufacturing industry, applications include predictive analytics for maintenance and repair of production lines, and analysis of consumer data to predict consumer buying patterns and fashion trends. Fashion retailers can use data analytics to improve the efficiency of their e-commerce business. Fashion tech company Virtusize enables online shoppers to buy the right size, either by measuring the clothes in their closet or by comparing specific brands and styles to their own (Virtusize 2021). Virtusize can increase average order values by 20% and decrease return rates by 30% by reducing uncertainty around size and fit. Similarly, online styling service Stitch Fix uses big data analytics to deliver personalized style recommendations to customers. All of Stich Fix's revenue results directly from its recommendations that combine data and machine learning with expert human judgment (Lake 2018).

A total of 52 textile and garment manufacturers in Punjab, Pakistan were surveyed for the employer surveys. The analysis shows that textile and garment manufacturers in Punjab, Pakistan have a limited grasp of 4IR technologies and their applications. Only 35% of firms surveyed indicated an intermediate or advanced level of understanding of 4IR technologies and the potential benefits that such technologies could bring to their industry (Figure 1). Only a quarter of firms surveyed indicated that firms in their supply chain currently use 4IR technologies or have plans to use them while 35% said that they did not know if firms in their supply chain use such technologies (Figure 2).

Interviews with local experts and stakeholders further suggest that firms and workers in the textile and garment manufacturing industry have a negative perception of 4IR technologies and associate automation with job losses. This could be due to the lack of a robust understanding of 4IR and therefore a clear perspective on the potential benefits that adoption of such technologies could bring to workers and the industry as a whole. This suggests a need for policies and programs to increase awareness of 4IR technologies and their potential benefits. Policies to support the adoption of 4IR technologies across the textile and garment manufacturing supply chain (i.e., suppliers, distributors, and retailers) would also need to be considered.

Most textile and garment manufacturing firms expect output per worker to increase between 10% and 25%. Overall, a 35% increase in output per worker is expected across the industry by 2025, with the adoption of 4IR technologies (Figure 3). There are estimates that the full adoption of 4IR technologies across the textile and garment manufacturing value chain could reduce labor time by around 40% to 70 (McKinsey & Company 2018). The more conservative estimates in Punjab can likely be attributed to generally low awareness of 4IR technologies.

Figure 1: Employers' Understanding of Industry 4.0 Technologies in the Textile and Garment Manufacturing Industry in Punjab, Pakistan

Only 35% of employers in the textile and garment manufacturing industry believe that they have a good understanding of 4IR technologies

Percent of surveyed firms

Novice: I have not heard of 4IR. — 19

Basic: I am aware of 4IR, but do not know its specific applications and their benefits to my company. — 46

Intermediate: I understand broadly what 4IR is, its application and benefits, but do not have a detailed understanding of how it can be deployed in my company. — 23

Advanced: I have a detailed understanding of 4IR and its applications, how it can be deployed, and its benefits for my company. — 12

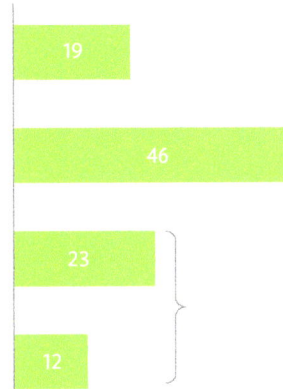

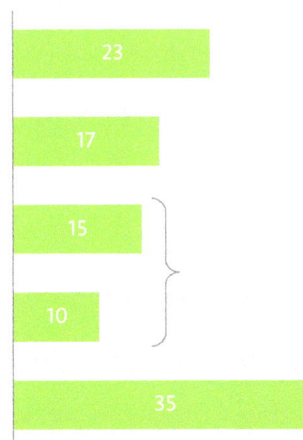

4IR = Fourth Industrial Revolution.
Note: Based on survey of employers in the textile and garment manufacturing industry between June and September 2021 (n=52).
Source: Asian Development Bank Sustainable Development and Climate Change Department.

Figure 2: Employers' Understanding of Industry 4.0 Technologies among Firms in the Textile and Garment Manufacturing Supply Chain in Punjab, Pakistan

Only a quarter of textile and garment manufacturers feel that companies in their supply chain have a good understanding of 4IR technologies

Percent of surveyed firms

Novice: Our supply chain companies have not heard of 4IR. — 23

Basic: Our supply chain companies are aware of 4IR technologies, but do not use them. — 17

Intermediate: Our supply chain companies have some understanding of 4IR technologies, and either use them on a small scale or plan to do so in the near future. — 15

Advanced: Our supply chain companies have a detailed understanding of 4IR technologies, and usethem in their operations. — 10

We do not know — 35

4IR = Fourth Industrial Revolution.
Note: Based on survey of employers in the textile and garment manufacturing industry between June and September 2021 (n=52).
Source: Asian Development Bank Sustainable Development and Climate Change Department.

Figure 3: Expected Increase in Output Per Worker Due to Industry 4.0 Technologies between 2020 and 2025 in the Textile and Garment Manufacturing Industry in Punjab, Pakistan

The majority of firms expect output per worker to increase by 10% to 25% in 5 years' time with the adoption of 4IR technologies

Percent of surveyed firms

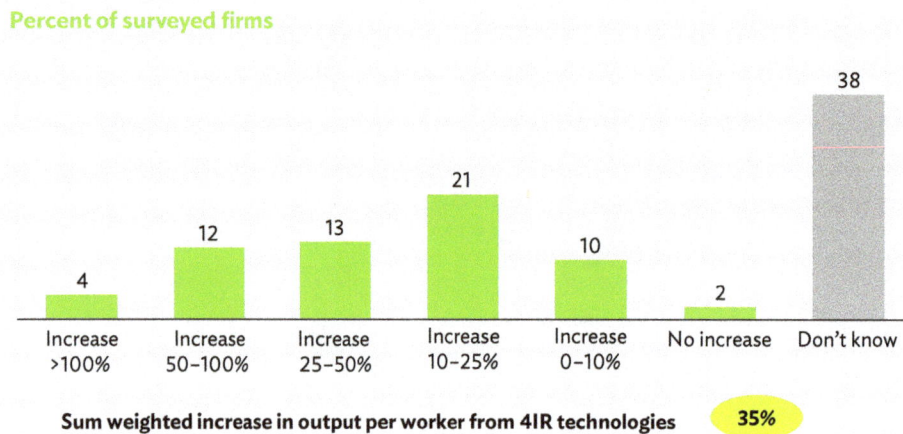

Sum weighted increase in output per worker from 4IR technologies   **35%**

4IR = Fourth Industrial Revolution.
Note: Based on survey of employers in the textile and garment manufacturing industry between June and September 2021 (n=52). Calculated using sum-weighted average of output increase by the number of firms indicating different levels of expected increase in output, i.e., 0%, 0%–10%, 10%–25%, 25%–50%, 50%–100%, and over 100%. The midpoint of the range for each option for expected increase in output is used; for expected output increase of over 100%, the lower bound of 100% is used.
Source: Asian Development Bank Sustainable Development and Climate Change Department.

Despite the limited understanding of 4IR technologies among textile and garment manufacturers, many in fact have plans to adopt 4IR technologies (Figure 4). This could suggest that while firms understand specific 4IR technologies and find them critical to their operations, they continue to see significant scope to expand their understanding of 4IR as a whole given its future importance. Some 27% of firms surveyed claimed that IOT technologies were already fully deployed across all possible functions in their firms, and only 8% of firms had not experimented with or used IOT technologies at all. At least 90% of firms plan to implement IOT, additive manufacturing, AI, and big data analytics technologies by 2025.

Only 27% of firms reported use of autonomous robots. This is despite the potential for automating up to 90% of sewing processes (ACT/EMP and ILO 2017). This is likely due to Pakistan's labor cost being relatively low compared to other textile and garment manufacturing countries and the ready availability of human resources (ILO 2016). Firms could therefore assess that the high capital investments needed to create fully automated manufacturing chains operated by autonomous robots would not pay off, opting instead to adopt partially automated supply chains with less sophisticated equipment. In addition, firms and workers in some developing economies might tend to associate the use of automated technologies with job losses and resist the adoption of such technologies (Thakur 2016). However, while 27% of firms have adopted autonomous robots as of 2020 (based on survey data in 2021), 52% of firms expect to use such technologies by 2025. This could be due to expectations that labor costs will increase in the future (ILO 2017). Artificial intelligence and IOT technologies are expected to be the most prevalent technologies in 2025.

Figure 4: Adoption of Relevant Industry 4.0 Technologies in the Textile and Garment Manufacturing Industry in Punjab, Pakistan

Deployment of autonomous robots lags other 4IR technologies today and in 5 years' time

Percent of surveyed firms

Extent of adoption by company

High[a]   Moderate[b]   Low[c]   None[d]

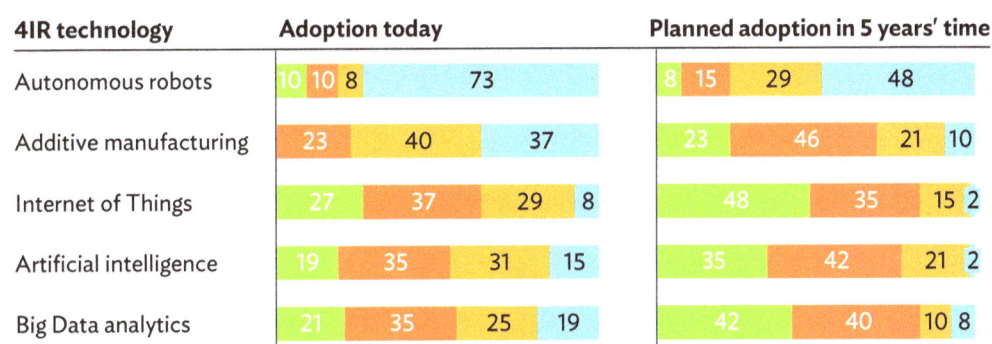

4IR technology	Adoption today				Planned adoption in 5 years' time			
Autonomous robots	10	10	8	73	8	15	29	48
Additive manufacturing		23	40	37	23	46	21	10
Internet of Things	27	37	29	8	48	35	15	2
Artificial intelligence	19	35	31	15	35	42	21	2
Big Data analytics	21	35	25	19	42	40	10	8

4IR = Fourth Industrial Revolution.
[a] "High:" Firm has fully deployed the technology across all possible functions in the enterprise and/or has plans to fully deploy the technology across all possible functions in the future.
[b] "Moderate:" Firm has implemented the technology, but not fully deployed across all possible functions in the enterprise and/or plans to implement the technology across a few functions in the future.
[c] "Low:" Firm is experimenting with the technology at a very limited scale within the enterprise and/or plans to experiment with the technology in the future.
[d] "None:" Firm has not used technology at all within the enterprise and/or has no plans to use the technology in the future.
Note: Based on survey of employers in the textile and garment manufacturing industry between June and September 2021 (n=52).
Source: Asian Development Bank Sustainable Development and Climate Change Department.

Past research on the impact of the pandemic on the textile and garment manufacturing industry points to a deepening divide in the industry. This is marked by the consolidation of firms focused on investing in advanced technologies and production methods on one end of the spectrum, and firms focused on large-scale, low-cost, and lower-skilled production at the other end (ILO 2020). The findings of the survey in Punjab reflect this divide with around half of firms agreeing that COVID-19 had accelerated the shift toward the adoption of 4IR technologies (Figure 5). This could lead to a widening gap between larger firms with ample resources and small and medium-sized enterprises (SMEs) in Punjab as the former would be able to reap significant productivity gains across the manufacturing value chain with the adoption of 4IR technologies.

Among the firms that agreed that COVID-19 had or will accelerate technology adoption, 74% cited the lack of labor due to movement restrictions as a key reason for the acceleration, as various lockdowns have been imposed in Punjab and nationwide since 2020. Other surveys show that 84% of textile and garment manufacturing firms in Pakistan were not operational during the height of the pandemic, out of which a major proportion was in Punjab.[9] The restrictions also accelerated the adoption of digital marketing channels by textile and garment

---

[9]   *Daily Times* 2021. Textile Industry of Pakistan in the Time of Covid-19. 16 March. https://go.gale.com/ps/i.do?p=HRCA&u=googlescholar&id=GALE|A655049241&v=2.1&it=r&sid=sitemap&asid=f6fc858d.

manufacturers in Pakistan as many turned to online platforms to market their products. Between July 2019 and June 2020, Pakistan's e-commerce market grew by 79% (*Xinhua* 2021). While companies had started investing in digital tools such as improved data analytics and demand forecasting even before the pandemic, COVID-19 increased the need to use precise digital tools to improve forecasting and supply change management. In addition, the disruption in international travel led to the use of tools such as virtual sampling and 3D product design tools (ILO 2020b). For instance, firms in Pakistan such as Masood Textile Mills and Amami Clothing started to use 3D garment simulation software CLO in their product development process. The software can create virtual prototypes and modelling to replace physical prototypes and reduce the overall time and cost of the design process.[10]

Figure 5: Perceptions on Impact of the COVID-19 Pandemic on Adoption of Industry 4.0 Technologies in the Textile and Garment Manufacturing Industry in Punjab, Pakistan

Around half of employers believe that the COVID-19 pandemic has accelerated or will accelerate the use of 4IR technologies

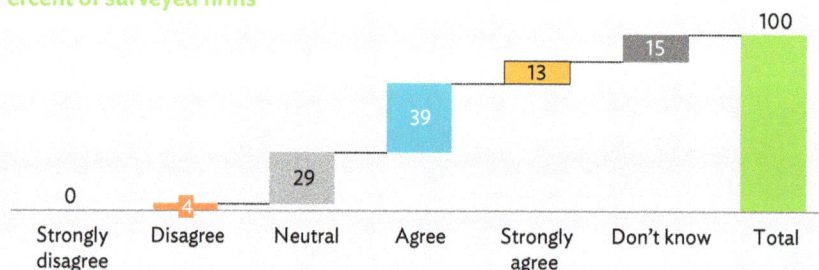

Percent of surveyed firms

	4	29	39	13	15	100
Strongly disagree	Disagree	Neutral	Agree	Strongly agree	Don't know	Total

Common reasons for accelerated adoption

Lack of labor due to movement restrictions necessitates more automation and shifting of activities to digital means

Strategic shift toward greater digitization by company's management

Note: Based on survey of employers in the textile and garment manufacturing industry between June and September 2021 (n=52).
Source: Asian Development Bank Sustainable Development and Climate Change Department.

Skills Demand Analysis

In carrying out the skills demand analysis, two parameters were evaluated: job implications and task implications. The study used data from the survey to consider the job implications of 4IR, by estimating job displacement and gains in the selected industries in each of the countries. For textile and garment manufacturing industry, 52 employers were surveyed; for the IT–BPO industry, 51 employers were surveyed; and for training institutions, 71 were surveyed.

---

[10]   Sources: local expert; Masood Textile. Production. https://masoodtextile.com/production/; and Amami. Production. https://amami.io/.

Job Implications

To determine the job implications of impact that adoption of 4IR technologies can have on employment in Punjab's textile and garment and manufacturing industry in 2025, the displacement and productivity effects of adopting 4IR technologies were estimated to determine the net change in job numbers (Figure 6).

(i) **Displacement effect.** This refers to the number of jobs that could be potentially lost due to automation using 4IR technologies. Around 31% or 1.13 million jobs out of the current workforce size in Punjab's textile and garment manufacturing industry could potentially be displaced due to the adoption of 4IR technologies between 2020 and 2025.

(ii) **Productivity effect.** This refers to the job gains due to improved productivity from technology adoption, which increases the potential total output of the industry. If policies that encourage full 4IR adoption are implemented, based on our estimates, up to 1.52 million new jobs could be created in Punjab's textile and garment manufacturing industry by 2025 due to the productivity effect.

This report estimates that the jobs created by the productivity effect exceed those displaced to create 390,000 new jobs or the equivalent of 11% of the 2020 textile and garment manufacturing workforce in net job gains by 2025. This is over and beyond the jobs created by BAU growth and the methodology is detailed in Box 1.

In interpreting the gains from 4IR adoption, it is important to understand that the 390,000 net jobs estimated will be created over and beyond new jobs due to BAU growth. In other words, the textile and garment manufacturing workforce grew at approximately 4.8% per annum[11] from 2015 and 2020. If this growth were extrapolated up to 2025, an additional 970,000 jobs would have been created even without adopting 4IR technologies. However, the full adoption of 4IR technologies could see up to 1.36 million more jobs in 2025 than in 2020.

There are, however, two caveats to realizing these job gains. First, these gains assume that firms in Punjab fully adopt 4IR technologies. However, only 35% of firms in the industry have a strong understanding of such technologies and their applications. Second, new jobs created will not be identical to the jobs displaced and could require different skill sets. In tandem with the shift in job scope of the average textile worker, the skills demanded by employers will also change. For instance, Lahore-based garment manufacturer, Combined Fabrics, adopted automated precision cutting machines that reduced the number of staff in the cutting room by 80% and allowed employees to be deployed to other tasks (Tukatech 2021). In Hong Kong, China, researchers developed WiseEye, an AI-based automated textile inspection system to help manufacturers instantly detect defects in fabrics during production. WiseEye can reduce loss and wastage due to faulty textiles by 90% as compared to human inspection (Del Buono 2018). The implementation of these 4IR technologies could substantively change the job scope of workers and potentially lead to the displacement of some workers who cannot adapt to new job roles. As such, strong retraining policies based on the new skill needs created by 4IR would be critical for workforces to reap the benefits of 4IR. Based on the employer surveys conducted, digital and ICT skills as well as adaptive learning capabilities will be increasingly sought by textile and garment manufacturing employers, and programs would need to be developed to ensure that workers have these capabilities.

---

[11] Calculated using industry and workforce data from Pakistan Bureau of Statistics.

Box 1: Estimating Employment Changes Due to Adoption of Industry 4.0 Technologies

To determine the impact that the adoption of technologies characteristic of the Fourth Industrial Revolution (Industry 4.0 or 4IR) will have on employment in Punjab's textile and garment and manufacturing industry from 2020 to 2025, data on the business-as-usual (BAU) growth scenario of the industry and responses from the employers' survey were utilized. The displacement and productivity effects of adopting 4IR technologies were estimated to determine the net change in jobs.

- **Displacement effect.** This refers to the number of jobs lost due to automation using 4IR technologies. It assumes that as output per worker increases due to the adoption of 4IR technologies, fewer workers would be needed to produce the same amount of output under a business-as-usual scenario in2025. The BAU amount of output in 2025 was calculated based on growth rates before the coronavirus disease (COVID-19) pandemic struck, i.e., from 2014 to 2019, while pre-COVID-19 industry labor productivity growth rates from 2014 and 2019 were used to calculate the BAU labor productivity (i.e., output per worker) for the next 5 years.[a] The expected labor productivity increase from the adoption of 4IR technologies, on top of the expected BAU labor productivity between 2020 and 2025, was obtained from the employer survey to calculate the displacement effect.

- **Productivity effect.** This refers to the job gains due to improved productivity from technology adoption, which increases the potential total output of the industry. For instance, the improved ease in production or the manufacturing of higher quality goods can both lead to higher total industry output and corresponding job creation. It assumes that (i) the market can completely absorb the higher output produced; and (ii) firms can produce at the original cost notwithstanding the higher productivity (e.g., pay wages based on the BAU productivity of workers prior to 4IR adoption). Using the labor productivity increase estimated from the survey, on top of the labor productivity increase that would have taken place at BAU by 2025, we calculated the industry's potential total output with 4IR adoption by 2025 and determined how many additional workers could be hired if labor productivity had remained at BAU levels in 2025.

- **Net gains.** The net gains were determined by taking the net of the increase in jobs created by the productivity impact and decrease in jobs created by the displacement impact.

[a] In Pakistan, national labor productivity growth rates were used as sectoral labor productivity growth rates were not available. Data was obtained from the Pakistan Bureau of Statistics.
Sources: Asian Development Bank Sustainable Development and Climate Change Department and Pakistan Bureau of Statistics.

To better understand the differences between the jobs displaced and the jobs created, it is important to recognize that each industry is characterized by various occupational roles carrying out different tasks, and the impact of 4IR technologies on these roles is uneven. For example, manual roles such as a sewing machine operator or weaver operating traditional hand-powered weaving machines typically face a higher risk of automation. Meanwhile, technology adoption could increase the number of technical roles as more workers are needed to operate and repair more advanced equipment or use digital platforms. In this study, jobs in the textile and garment manufacturing industry are split into five occupational groups (Table 2).

Figure 6: Estimated Impact of Industry 4.0 on Number of Jobs between 2020 and 2025 in the Textile and Garment Manufacturing Industry in Punjab, Pakistan

The adoption of 4IR technologies will lead to 11% more jobs in 5 years' time as the number of newly created jobs outweigh displaced jobs

Percent of jobs impacted due to displacement and productivity effects of 4IR in 5 years' time

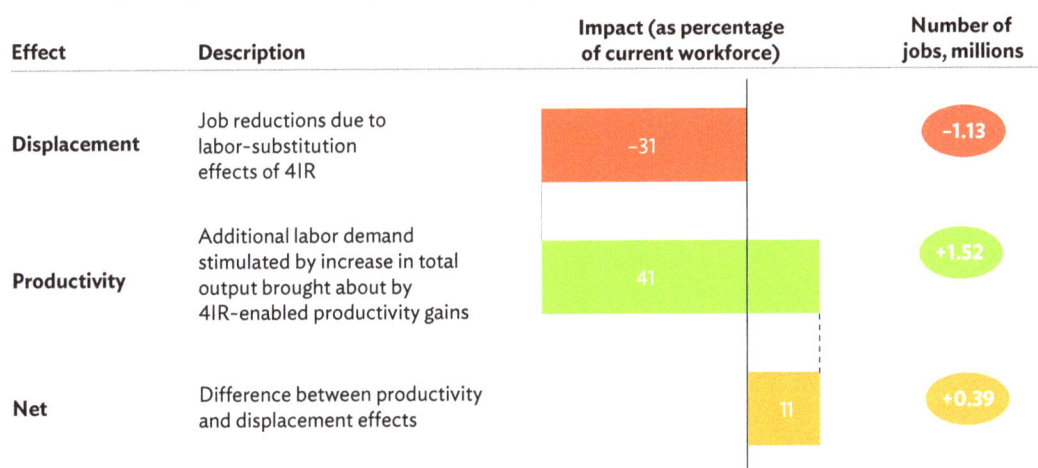

Effect	Description	Impact (as percentage of current workforce)	Number of jobs, millions
Displacement	Job reductions due to labor-substitution effects of 4IR	–31	–1.13
Productivity	Additional labor demand stimulated by increase in total output brought about by 4IR-enabled productivity gains	41	+1.52
Net	Difference between productivity and displacement effects	11	+0.39

4IR = Fourth Industrial Revolution
Notes: Based on survey of employers in the textile and garment manufacturing industry between June and September 2021 (n=52). Industry employment data and gross domestic output data are from the Pakistan Bureau of Statistics.
Sources: Asian Development Bank Sustainable Development and Climate Change Department and Pakistan Bureau of Statistics.

Table 2: Occupational Groups in the Textile and Garment Manufacturing Industry

	Occupational Group	Possible Job Titles
1	**Technical**	• Engineering technician • Garment technologist • Sewing production line operator
2	**Managerial**	• Chief executive officer • Factory floor manager
3	**Customer-facing**	• Marketing manager • Retail sales associate
4	**Administrative**	• Secretary • Finance executive
5	**Elementary and/or manual jobs**	• Sewing machine operator • Weaver

Source: AlphaBeta.

Overall, our analysis shows that manual occupations are more likely to be displaced by the adoption of 4IR technologies compared to other occupations (Figure 7). Some 17% of employers expect the number of manual jobs to decline by more than half while 25% expect a more moderate decrease. In contrast, firms expect the adoption of 4IR technologies to create more technical jobs.

Figure 7: Employers' Expectations on Impact of Industry 4.0 on the Number of Jobs between 2020 and 2025 in the Textile and Garment Manufacturing Industry in Punjab, Pakistan

In the textile and garment manufacturing industry, manual jobs are expected to decrease most significantly due to 4IR technologies adoption

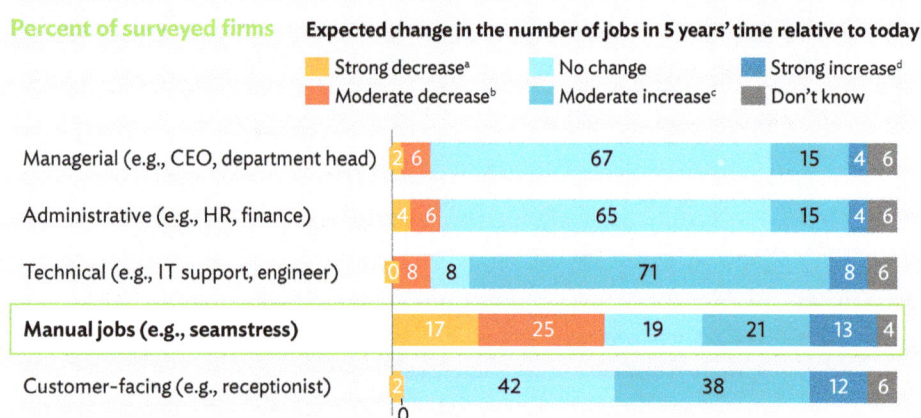

Percent of surveyed firms   Expected change in the number of jobs in 5 years' time relative to today

Strong decrease[a]   No change   Strong increase[d]
Moderate decrease[b]   Moderate increase[c]   Don't know

Occupation	Strong decrease	Moderate decrease	No change	Moderate increase	Strong increase	Don't know
Managerial (e.g., CEO, department head)	2	6	67	15	4	6
Administrative (e.g., HR, finance)	4	6	65	15	4	6
Technical (e.g., IT support, engineer)	0	8	8	71	8	6
Manual jobs (e.g., seamstress)	17	25	19	21	13	4
Customer-facing (e.g., receptionist)	2		42	38	12	6

CEO = chief executive officer, HR = human resources, IT = information technology.
[a] Greater than or equal to 50% decrease in number of jobs.
[b] Less than 50% decrease in number of jobs.
[c] Less than 50% increase in number of jobs.
[d] Greater than or equal to 50% increase in number of jobs.
Note: Based on survey of employers in the textile and garment manufacturing industry between June and September 2021 (n=52).
Source: Asian Development Bank Sustainable Development and Climate Change Department.

These shifts will change the composition of jobs in the industry (Figure 8). Manual occupations are 40.6% of all jobs in 2020 but are expected to become 36.6% of all jobs by 2025. In contrast, the proportion of technical roles is expected to increase over the same period. While the shift in workforce composition is based on employers' projections of the impact of 4IR on the number of jobs in each occupational group and could be limited by a generally poor understanding of 4IR, a gradual shift toward fewer manual jobs and more technical roles is nevertheless observed. The manual occupations displaced are likely to have significantly different skill profiles from the technical occupations created, so that displaced workers would not be able to transition seamlessly across roles without retraining. The challenge is compounded by the fact that workers engaged in manual work are likely to be poorly educated or illiterate and would have difficulties finding alternative work or seeking training. In Punjab, 39% of the working population is illiterate and cannot meet the educational qualification requirements of formal skills training providers to pursue further training (ADB 2020). As such, it would be critical to build strong safety nets to support the reskilling needs of these workers before they are made redundant or obsolete by the job shifts created by the adoption of 4IR technologies, and to ensure that they too can benefit from the

Figure 8: Composition of Jobs between 2020 and 2025 in the Textile and Garment Manufacturing Industry in Punjab, Pakistan

The distribution of jobs will change—manual jobs will record the largest fall while technical jobs will see the largest increase in 5 years' time

Weighted average percentage share of employees by occupational group in surveyed firms[a]

Legend: 🟡 Negative shift   🟢 Positive shift

Occupational group	Share today	Share in 5 years' time[b]	Percentage shift
Manual jobs	40.6	36.6	-4.0%
Administrative	12.3	11.9	-0.4%
Managerial	10.6	11.4	+0.8%
Customer-facing	7.6	8.8	+1.2%
Technical	28.9	31.2	+2.3%

[a] Average share of employees in surveyed firms is weighted by the number of employees in each firm, as indicated by respondents; percentages might not add up to 100% due to rounding.

[b] The change in the number of workers in each job type is based on the number of firms indicating different levels of changes in number of jobs, i.e., "strong increase," "moderate increase," "no change," "moderate decrease," "strong decrease." The midpoint of the range for each option. For expected change is used; for expected increase or decrease of over 50%, the low bound of 50% was used.

Note: Based on survey of employers in the textile and garment manufacturing industry between June and September 2021 (n=52).

Source: Asian Development Bank Sustainable Development and Climate Change Department.

job gains created by 4IR. For instance, under the Malaysian Skills Certification Program, skill certificates are granted to workers who do not have any formal educational qualifications but have obtained relevant knowledge, experience, and skills in the workplace to allow them to seek professional training and enhance their career prospects.[12] International agencies such as ADB could lend their financial and technical support in this area.

Apart from ensuring that less educated manual workers can benefit from the shift toward 4IR, it is equally important that women are not left behind in these gains. Male workers account for close to two thirds of all textile and garment workers in Pakistan and most new jobs created in the industry will be gained by male workers, as shown in Figure 9 (ILO 2019a). In addition, most of the jobs created in the industry will be in technical roles, in which the concentration of male workers is particularly high. Industry 4.0 could thus exacerbate inequalities within the labor market if strong reskilling policies for female workers are not implemented.

---

[12]   Department of Skills Department. Malaysian Skill Certificate (SKM). https://www.dsd.gov.my/index.php/en/service/malaysian-skills-certificate.

Figure 9: Estimated Net Job Gains by Gender from Industry 4.0 Adoption between 2020 and 2025 in the Textile and Garment Manufacturing Industry in Punjab, Pakistan

The net gains in jobs will benefit male workers more than female workers, implying that policies are needed to ensure that adoption benefits are equitable

Estimated number of net jobs created by gender (in thousands)

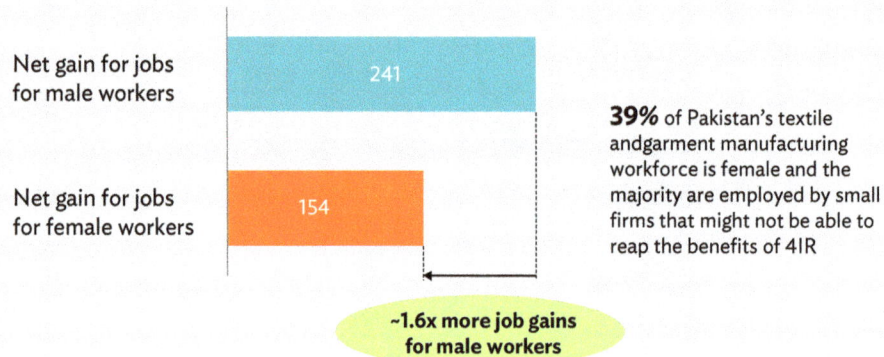

Net gain for jobs for male workers — 241

Net gain for jobs for female workers — 154

39% of Pakistan's textile andgarment manufacturing workforce is female and the majority are employed by small firms that might not be able to reap the benefits of 4IR

~1.6x more job gains for male workers

4IR = Fourth Industrial Revolution

Note: Based on survey of employers in the textile and garment manufacturing industry between June and September 2021 (n=52). Industry employment data and gross domestic output data are from the Pakistan Bureau of Statistics.

Sources: Asian Development Bank Sustainable Development and Climate Change Department and Pakistan Bureau of Statistics.

Task Implications

To understand the impact of the adoption of 4IR technologies on jobs and skills demand, it is important to understand that technology does not automate jobs, but rather individual tasks or combinations of tasks. This research examines five types of tasks linked to jobs in the textile and garment manufacturing industry and how they could be impacted by 4IR:

(i)   **Routine physical.** These tasks involve repetitive and predictable physical work. For example, a seamstress creating a handsewn traditional garment.

(ii)   **Routine interpersonal.** These tasks involve predictable interactions with other people. For example, a retail sales assistant serving a customer.

(iii) **Nonroutine physical.** These tasks involve physical work that is not repetitive or predictable. For example, a mechanic diagnosing and repairing factory equipment.

(iv)   **Nonroutine interpersonal.** These tasks involve complex or creative interactions with other people. For example, supervising others or making speeches or presentations.

(v)   **Analytical.** These are tasks that vary significantly and involve a strong thinking and analytical component. They predominantly involve computers or other technological equipment.

Employers were asked to estimate how much time an average employee spent on each task type in an average work week as of 2020 (based on 2021 survey data) and predict how that will change by 2025. The analysis revealed a slight drop in the proportion of time spent on routine tasks with the time saved reallocated to analytical and nonroutine tasks. The proportion of time spent on routine physical tasks is expected to fall from 55.7% to 52.3% (Figure 10). Past research shows that technology adoption could lead to an increase in work satisfaction as machines take over a greater share of full routine tasks as the monotonous, automatable tasks performed by typically low-skilled workers are also the least satisfying tasks to perform (AlphaBeta 2017).

Figure 10: Time Spent by Employees on Tasks at Work between 2020 and 2025 in the Textile and Garment Manufacturing Industry in Punjab, Pakistan

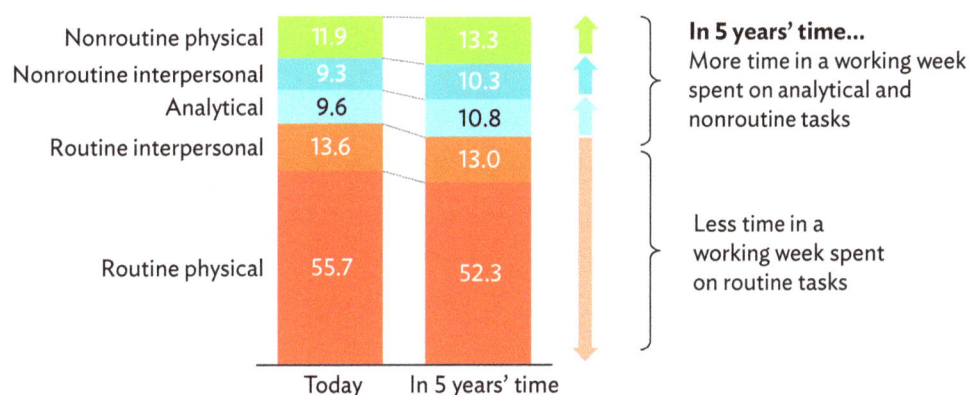

Adoption of 4IR technologies may shift the distribution of weekly working hours toward analytical and nonroutine tasks

Average percentage share of weekly working hours spent by task in surveyed firms

	Today	In 5 years' time
Nonroutine physical	11.9	13.3
Nonroutine interpersonal	9.3	10.3
Analytical	9.6	10.8
Routine interpersonal	13.6	13.0
Routine physical	55.7	52.3

In 5 years' time...
More time in a working week spent on analytical and nonroutine tasks

Less time in a working week spent on routine tasks

Notes: Based on survey of employers in the textile and garment manufacturing industry between June and September 2021 (n=52).
Source: Asian Development Bank Sustainable Development and Climate Change Department.

Skills Implications

These tasks shifts will impact the skills required in the industry. While employers indicated only slight shifts in the proportion of time allotted to each type of tasks, the skills that are valued by employers in the textile and garment manufacturing industry will see a change in relative importance due to these shifts. This analysis considers 10 categories of skills as set out in Table 3.

Table 3: Categories of Skills Considered in the Analysis

No.	Skill	Definition
1	**Creative thinking and/or design**	Ability to develop, design, or creating new applications, ideas, relationships, systems, or products
2	**Critical thinking**	Ability to use logic and reasoning to identify the strengths and weaknesses of alternative solutions, conclusions or approaches to problems
3	**Adaptive learning**	Ability to pick up new skills as demanded by the job
4	**Complex problem solving**	Ability to identify complex problems and review related information to develop and evaluate options and implement solutions
5	**Digital and/or ICT skills**	Ability to design, set-up, operate, and correct malfunctions involving application of machines or technological systems
6	**Numeracy**	Ability to add, subtract, multiply, or divide quickly and correctly and use mathematics to solve problems
7	**Written communication**	Ability to read and understand information and ideas presented in writing, and to communicate information and ideas in writing
8	**Verbal communication**	Ability to communicate information and ideas clearly by talking to others
9	**Management**	Ability to motivate, develop, and direct people as they work, and to identify the best people for the job
10	**Social and interpersonal**	Ability to work with people to achieve goals

ICT = information and communication technology.
Source: Asian Development Bank Sustainable Development and Climate Change Department.

The analysis reveals significant changes in the skills sought by employers in the textile and garment manufacturing industry in 2025, compared to what is valued by employers in 2020.

Digital and ICT skills are expected to become more important for employers by 2025 while written and verbal communication as well as social and interpersonal skills are expected to become less valued (Figure 11). Apart from digital and ICT skills, critical thinking and adaptive learning skills will also be sought after by employers in the future. The rising importance of these skills is consistent with the shift of the industry toward 4IR adoption. More workers are likely expected to operate simple machinery or ICT-based equipment as 4IR technologies are adopted in various parts of the manufacturing value chain.

While digital and ICT skills were ranked top by employers in terms of relative importance by 2025, reskilling needs in this area are relatively low. Instead, creative thinking or design as well as verbal and written communication are where employers see the largest proficiency gaps in 2020 (Figure 12). Creative thinking and critical thinking could be prioritized as areas to reskill employees, given that they are important to employers in 2025 and see significant skill gaps. Of the employers that indicated that a step-up from basic proficiency in creative thinking or design skills is needed, 29% stated that a step-up to the intermediate level of proficiency is needed while 71% would like to see a step-up to advanced proficiency.

Figure 11: Importance of Skills in 2020 and for Industry 4.0 Adoption by 2025 in the Textile and Garment Manufacturing Industry in Punjab, Pakistan

There is a significant change in skills perceived as important by employers over the next 5 years

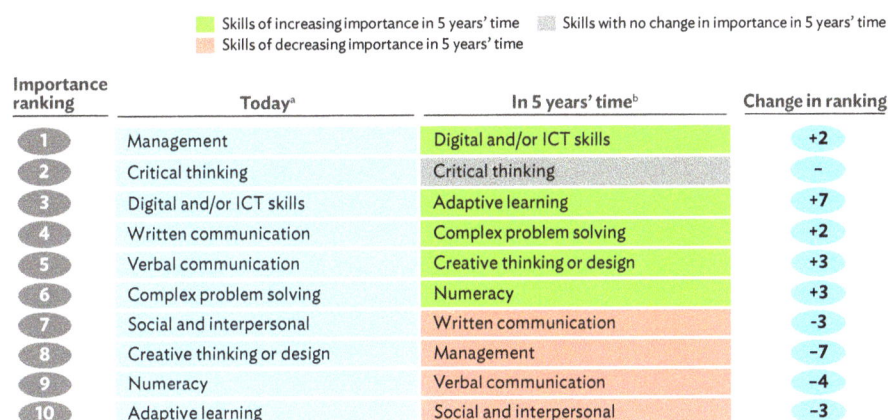

Skills of increasing importance in 5 years' time   Skills with no change in importance in 5 years' time
Skills of decreasing importance in 5 years' time

Importance ranking	Today[a]	In 5 years' time[b]	Change in ranking
1	Management	Digital and/or ICT skills	+2
2	Critical thinking	Critical thinking	–
3	Digital and/or ICT skills	Adaptive learning	+7
4	Written communication	Complex problem solving	+2
5	Verbal communication	Creative thinking or design	+3
6	Complex problem solving	Numeracy	+3
7	Social and interpersonal	Written communication	-3
8	Creative thinking or design	Management	-7
9	Numeracy	Verbal communication	-4
10	Adaptive learning	Social and interpersonal	-3

ICT = information and communication technology.

[a] Evaluated using employer survey and supported by job portal data.

[b] Evaluated using the employer survey.

Notes: Based on survey of employers in the textile and garment manufacturing industry between June and September 2021 (n=52); job data on the textile and garment manufacturing industry from the job portal Rozee (accessed 11 June 2021).

Source: Asian Development Bank Sustainable Development and Climate Change Department.

Figure 12: Required Step-Up in Level of Proficiency of Employees' Skills for Industry 4.0 Adoption between 2020 and 2025 in the Textile and Garment Manufacturing Industry in Punjab, Pakistan

To be 4IR-ready, workers would require proficiency leaps in creative thinking as well as written and verbal communication skills

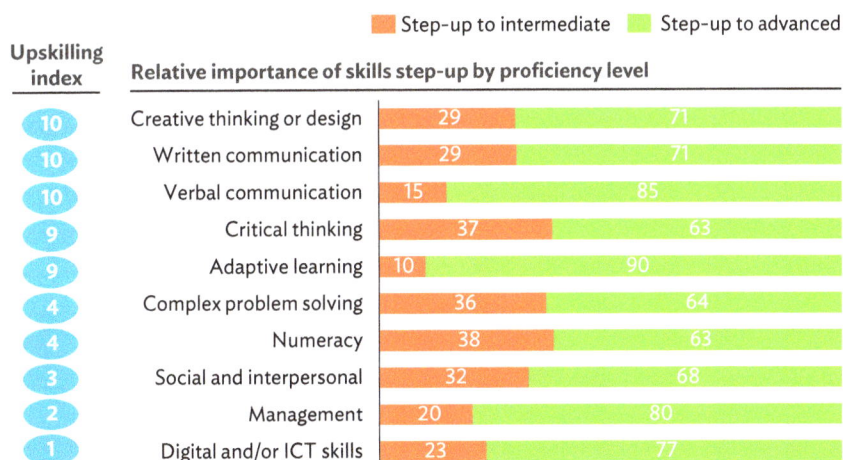

Step-up to intermediate   Step-up to advanced

Upskilling index	Relative importance of skills step-up by proficiency level		
10	Creative thinking or design	29	71
10	Written communication	29	71
10	Verbal communication	15	85
9	Critical thinking	37	63
9	Adaptive learning	10	90
4	Complex problem solving	36	64
4	Numeracy	38	63
3	Social and interpersonal	32	68
2	Management	20	80
1	Digital and/or ICT skills	23	77

ICT = information and communication technology.

Notes: Based on survey of employers in the textile and garment manufacturing industry between June and September 2021 (n=52). Index based on the number of employers indicating a need for workers with basic proficiency to be upskilled for each skill.

Source: Asian Development Bank Sustainable Development and Climate Change Department.

Skills Supply Trends

Of the employers surveyed, 50% strongly agreed that there is a large variance in the quality of graduates available for hire but 73% strongly believed that there are sufficient graduates to meet their firm's entry-level hiring needs (Figure 13). In contrast, a similar study conducted in Cambodia reported that only 35% of companies believed there were sufficient graduates to meet their firm's entry-level hiring needs (ADB 2021). The strong confidence in graduate quality could be due to a combination of entry-level skills demanded in the industry being relatively basic, as well as the willingness of employers in Punjab to provide training. Meanwhile, 67% of firms surveyed strongly agreed that their employees received an adequate amount of training and over half strongly agreed that investment in worker training was sufficient (Figure 14). While a large proportion of textile and garment manufacturing firms believe that they provide sufficient training to workers, expert interviews in Pakistan suggest that more structured training programs for new workers, particularly those without previous training or experience, carried out in well-equipped training and development centers, could further raise the overall quality of the workforce. In particular, it would be critical to provide targeted training to different groups of workers with different capabilities and scope of work (i.e., a manager vs. a factory floor worker, an experienced hire vs. a fresh hire) and training programs would need to be suitably tailored.

Figure 13: Employer Sentiment Toward Graduates Hired in the Textile and Garment Manufacturing Industry in Punjab, Pakistan

About 73% of employers strongly believe that there are sufficient graduates to meet their company's entry-level hiring needs

Percent of surveyed firms

Legend: Strongly agree | Neither agree nor disagree | Don't know or not applicable | Agree | Disagree or strongly disagree

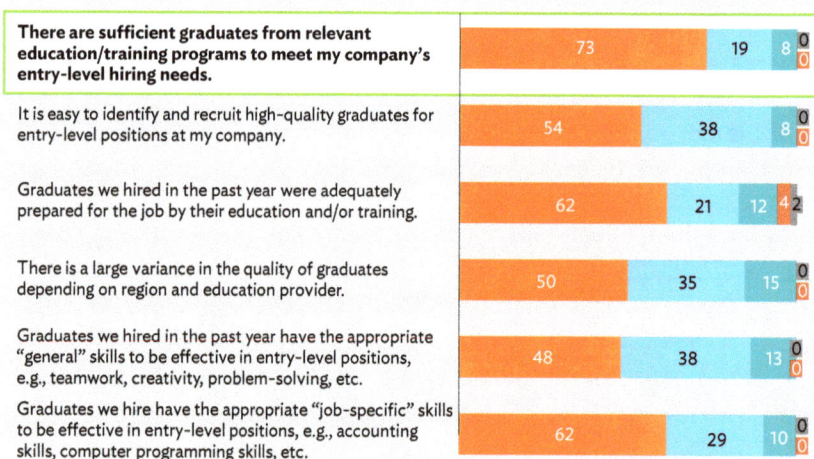

Statement	Strongly agree	Agree	Neither agree nor disagree	Disagree or strongly disagree	Don't know or not applicable
There are sufficient graduates from relevant education/training programs to meet my company's entry-level hiring needs.	73	19	8	0	0
It is easy to identify and recruit high-quality graduates for entry-level positions at my company.	54	38	8	0	0
Graduates we hired in the past year were adequately prepared for the job by their education and/or training.	62	21	12	4	2
There is a large variance in the quality of graduates depending on region and education provider.	50	35	15	0	0
Graduates we hired in the past year have the appropriate "general" skills to be effective in entry-level positions, e.g., teamwork, creativity, problem-solving, etc.	48	38	13	0	0
Graduates we hire have the appropriate "job-specific" skills to be effective in entry-level positions, e.g., accounting skills, computer programming skills, etc.	62	29	10	0	0

Note: Based on survey of employers in the textile and garment manufacturing industry between June and September 2021 (n=52).
Source: Asian Development Bank Sustainable Development and Climate Change Department.

Figure 14: Employers' Perception on Training for Employees in the Textile and Garment Manufacturing Industry in Punjab, Pakistan

About 67% of surveyed employers strongly believe that their employees receive an adequate amount of training to do their jobs well

Percent of surveyed firms

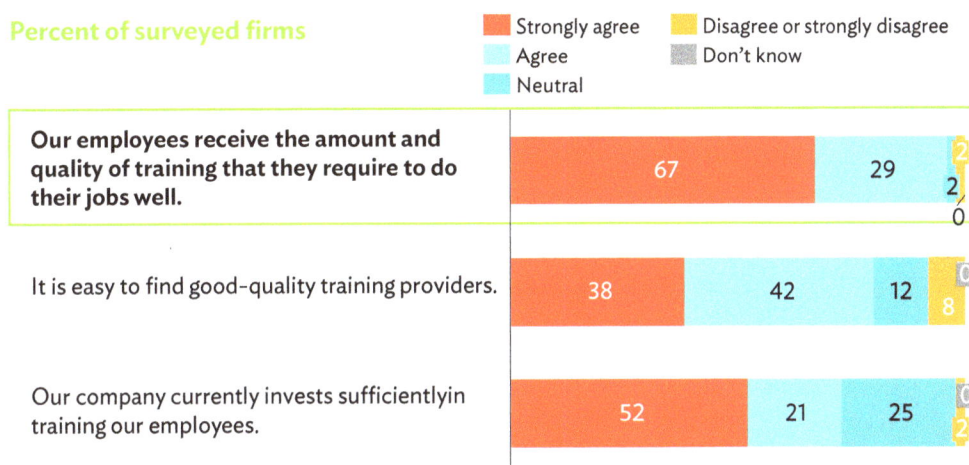

Legend:
- Strongly agree
- Agree
- Neutral
- Disagree or strongly disagree
- Don't know

Statement	Strongly agree	Agree	Neutral	Disagree or strongly disagree	Don't know
Our employees receive the amount and quality of training that they require to do their jobs well.	67	29		2 / 2	0
It is easy to find good-quality training providers.	38	42	12	8	0
Our company currently invests sufficiently in training our employees.	52	21	25	2	0

Note: Based on survey of employers in the textile and garment manufacturing industry between June and September 2021 (n=52).
Source: Asian Development Bank Sustainable Development and Climate Change Department.

To understand the types of training channels that employers relied on to address the skills shortage, four types of training channels were examined in relation to how they would be tapped to provide skills training for employees in 2020 and going forward (Table 4).

Table 4: Four Types of Training Channels

	Training Channel	Description
1	**On-the-job training**	Training that takes place within the firm as the employee performs the actual work. These are typically provided by a more senior or experienced coworker and can also be in the form of internally organized sessions conducted by coworkers.
2	**Flexible online training**	Online courses that are subsidized or sponsored by the firm for their employees (e.g., courses from online training platforms like Udemy and Coursera). Such online training courses tend to be flexible in terms of when workers may access the content, and typically allow workers to gain industry-recognized micro-credentials or micro-degrees at the end of the course.
3	**Professional courses**	Short courses that are sponsored or organized by the firm for their employees. These are conducted by professional instructors and are typically held within contained periods spanning at least 1 week and up to 6 months.
4	**Formal education courses**	Such courses are those taken at higher education or technical and vocational education and training institutions that are subsidized or sponsored by the firm for their employees. Such courses tend to be specially designed for working professionals, such as part-time diplomas or master's degrees.

Source: Asian Development Bank Sustainable Development and Climate Change Department.

On-the-job training appears to be the preferred training channel for most employers as of 2020 (based on survey conducted in 2021), and this will remain unchanged by 2025. In July 2021, Punjab became the first province in Pakistan to pass a new Apprenticeship Act to simplify approvals to register apprenticeship programs and facilitate stronger collaboration between training institutions and employers in providing on-the-job training supplemented by technical and vocational education in the classroom (*Dawn News* 2021). Interviews with local experts similarly suggest that while on-the-job training is prevalent in Pakistan's textile and garment manufacturing industry, it could be made more effective if combined with theoretical training that would expose workers to best practices abroad as well as technologies not currently used in Pakistan. Stronger collaboration between industry and training institutions to put in place industry apprenticeship and workplace-based training frameworks could help to achieve this.

Figure 15: Proportion of Employees Receiving Training in 2020 and Requiring Training by 2025 in Each Training Channel in the Textile and Garment Manufacturing Industry in Punjab, Pakistan

Employers believe that more training needs to take place across all channels, particularly flexible online training and professional courses

Percentage share of employees by training channel

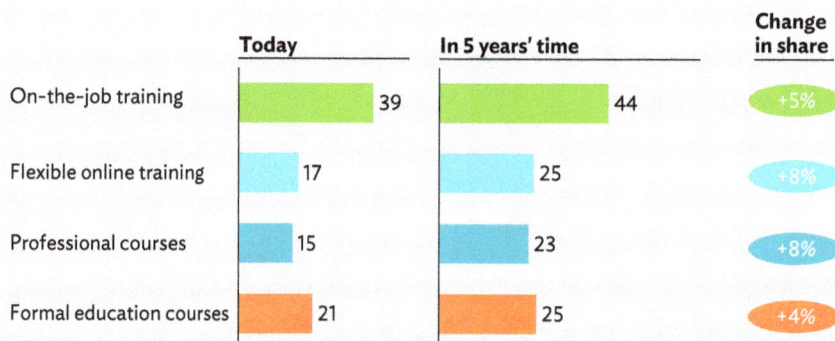

	Today	In 5 years' time	Change in share
On-the-job training	39	44	+5%
Flexible online training	17	25	+8%
Professional courses	15	23	+8%
Formal education courses	21	25	+4%

Notes: Based on survey of employers in the textile and garment manufacturing industry between June and September 2021 (n=52). The sum of all shares for 2020 and by 2025 exceed 100%, as one employee can undergo training in more than one training channel.
Source: Asian Development Bank Sustainable Development and Climate Change Department.

D. Information Technology–Business Process Outsourcing Industry

As expected, firms in the IT–BPO industry in Punjab demonstrate a better understanding of 4IR technologies and applications compared to textile and garment manufacturing firms. More than 50% of firms surveyed indicated that they had a strong grasp of 4IR technologies, up from 35% in the textile and garment manufacturing industry. Correspondingly, expectations of productivity gains are also more optimistic. Overall, IT–BPO firms expect to see output per worker increase by 45% by 2025 due to 4IR technologies. This is 10 percentage points higher than the gains estimated by textile and garment manufacturers. Meanwhile, 84% of IT–BPO firms surveyed agree that the COVID-19 pandemic has accelerated or will accelerate the adoption of 4IR technologies in their field.

As in the textile and garment manufacturing industry, the full adoption of 4IR technologies in the IT–BPO sector is expected to create net job gains. Close to 7,000 new jobs or the equivalent of 18% of the 2020 IT–BPO workforce are expected to be created. This gain is over and beyond the BAU growth of the industry's labor force. Technical occupations already form the largest proportion of jobs in the industry as of 2020, but the share of technical jobs is expected to further widen with the adoption of 4IR technologies. Skills such as creative thinking and the ability to solve complex problems will be increasingly demanded by employers. The relative importance of digital and/or ICT skills will fall, likely because employers would expect the bulk of the IT–BPO workforce to already have such skills and look instead for harder-to-find skill sets in workers.

Relevance to Industry 4.0

A range of 4IR technologies have the potential to transform the services provided by the IT–BPO industry and improve productivity as well as customer experience. Some of these key technologies are:

(i) **Artificial intelligence.** AI gives machines the ability to learn and act intelligently and carry out a wide range of human-like processes. This means they can make decisions, carry out tasks, and even predict future outcomes based on what they learn from the data. AI-enabled chatbots can be used as artificial agents that converse with customers and provide advice when live agents are unavailable. Past research has shown that health care firms and banks that use chatbots to deal with customer queries can save around 4 minutes, or more than $0.50 per inquiry (Olson 2018). Businesses can reduce customer service costs by up to 30% by implementing conversational solutions like virtual agents and chatbots (Reddy 2017).

(ii) **Systems integration.** Systems integration technologies link different computing systems and software applications to act as a coordinated whole. These include computer networking systems, enterprise application integration, and the integration of data management systems. Research by McKinsey shows that improvements in data sourcing, architecture, governance, and consumption can cut annual data spend by 5%–15% and put high-quality data within easier reach (McKinsey Digital 2020). Pakistan's Infotech is a provider of integrated systems (InfoTech 2021). One of its products is "Capizar," a capital markets product suite that includes automated trading platforms, electronic depository systems, regulatory compliance portals and market surveillance systems. "Capizar" has been implemented in 13 markets and processes more than a billion dollars' worth of equities and bonds trades daily (Pro Pakistani 2015).

(iii) **Augmented reality.** Augmented reality or AR creates an interactive experience of a real-world environment where real-world objects are enhanced by computer-generated perceptual information, sometimes across multiple sensory modalities. For instance, technical support teams at IT–BPO centers can leverage AR so that technicians can see a customer's surroundings and provide remote assistance such as for computer hardware repairs. Nestlé uses AR to provide remote support to its production and research and development (R&D) sites globally (Nestlé 2020). Through these digital tools, technical experts can support multiple projects at the same time, increasing efficiency. Studies have found that providing remote assistance through AR-based customer support can increase first call resolution by 20% and decrease the need for technician dispatch by 17% (Shaham 2020).

(iv) **Cloud computing.** Cloud computing is on-demand access, via the internet, to computing resources, including applications, data storage, and networking capabilities hosted at a remote data center and managed by a cloud service provider. The three main types of cloud computing services are Infrastructure-as-a-Service (IaaS), Platforms-as-a-Service (PaaS), and Software-as-a-Service (SaaS). IaaS allows customers to rent infrastructure on demand without worrying about data center maintenance, real estate, and rentals. PaaS allows applications to be rented on variable pricing models with limited upfront capital expenditures. SaaS provides precreated and tested application components that reduces or removes the need for firms to develop commonly used software functionality from scratch. For IT–BPO firms, the on-demand access and pay-per-use model for cloud computing services allows them to scale their operations up or down quickly without impacting operational efficiency or incurring high capital costs. Cloud computing services can also bring down processing time for data-intensive business processes so that IT–BPO firms can offer data processing workflow, from documentation and image processing to publishing, at a reduced turnaround time.[13] Research by BCG Platinion estimates that cloud computing can reduce overall IT spend by as much as 10%. In particular, the use of SaaS or web-based software applications such as Google Workspace can lower computing costs for end users by up to 35% and cut the risk of data breaches by more than 95% (*Google Cloud Blog* 2021). In Pakistan, software firms such as AVOXI and Intellicon offer cloud-based call center software solutions (*Intellicon 2017,* AVOXI 2021).

(v)   **Cybersecurity.** Cybersecurity refers to the protection of internet-connected systems from unauthorized exploitation (i.e., cyberattacks). In the IT–BPO industry, cybersecurity technology can help to manage threats to internal databases that contain sensitive client information. The provision of cybersecurity services is a potential source of revenue for IT–BPO firms. Research commissioned by Skurio, a provider of digital risk protection solutions, shows that over 50% of businesses in the United Kingdom are turning to outsourced partners for their cybersecurity needs, citing the lack of in-house expertise as one of the key reasons (*Skurio* 2020).

Figure 16: Understanding of Industry 4.0 Technologies in the Information Technology–Business Process Outsourcing Industry in Punjab, Pakistan

About 45% of employers in the IT-BPO industry have an intermediate understanding of 4IR technologies and their applications

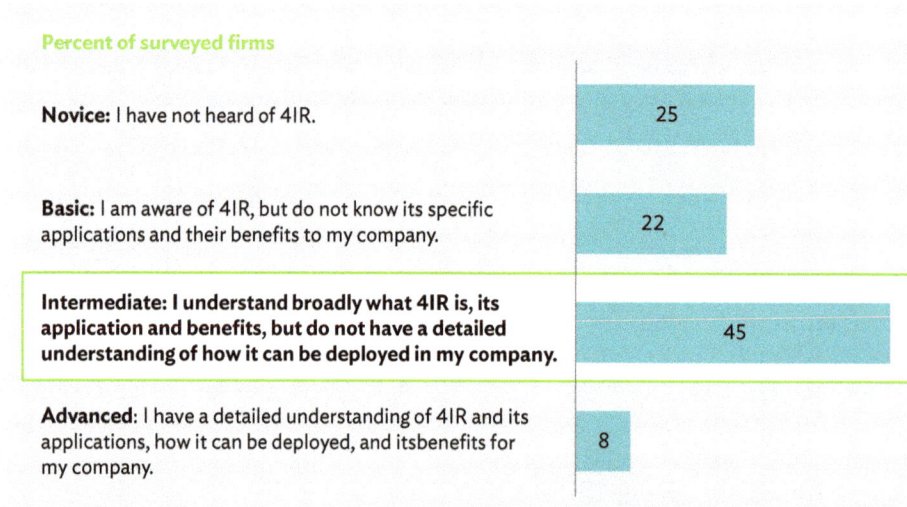

Percent of surveyed firms

Novice: I have not heard of 4IR.	25
Basic: I am aware of 4IR, but do not know its specific applications and their benefits to my company.	22
Intermediate: I understand broadly what 4IR is, its application and benefits, but do not have a detailed understanding of how it can be deployed in my company.	45
Advanced: I have a detailed understanding of 4IR and its applications, how it can be deployed, and itsbenefits for my company.	8

4IR = Fourth Industrial Revolution, IT-BPO = information technology-business process outsourcing.
Note: Based on survey of employers in the IT–BPO industry between June and September 2021 (n=51).
Source: Asian Development Bank Sustainable Development and Climate Change Department.

---

[13]   WNS. Does the Cloud Come with a Silver Lining for BPO? https://www.wns.com/insights/articles/articledetail/81/does-the-cloud-come-with-a-silver-lining-for-bpo.

For the IT–BPO employer surveys, 51 firms surveyed in Punjab. Unsurprisingly, IT–BPO firms surveyed demonstrated stronger understanding of 4IR technologies compared to firms in the textile and garment manufacturing industry (Figure 16). Similarly, half of IT–BPO firms surveyed in Punjab believe that firms in their supply chain understand 4IR technologies well and only 12% said that firms in their supply chain have not heard of 4IR technologies (Figure 17). The stronger understanding is reflected also in the assessment of potential labor productivity gains, as 35% of firms expect output per worker to increase by 50%–100% between 2020 and 2025 with the adoption of such technologies. The overall increase in output per worker across the industry is estimated to be 45%, 10 percentage points higher than for the textile and garment manufacturing industry (Figure 18).

Figure 17: Understanding of Industry 4.0 Technologies Among Firms in the Information Technology–Business Process Outsourcing Supply Chain in Punjab, Pakistan

Only 12% of IT-BPO firms believe that companies in their supply chain have not heard of 4IR technologies

Percent of surveyed firms

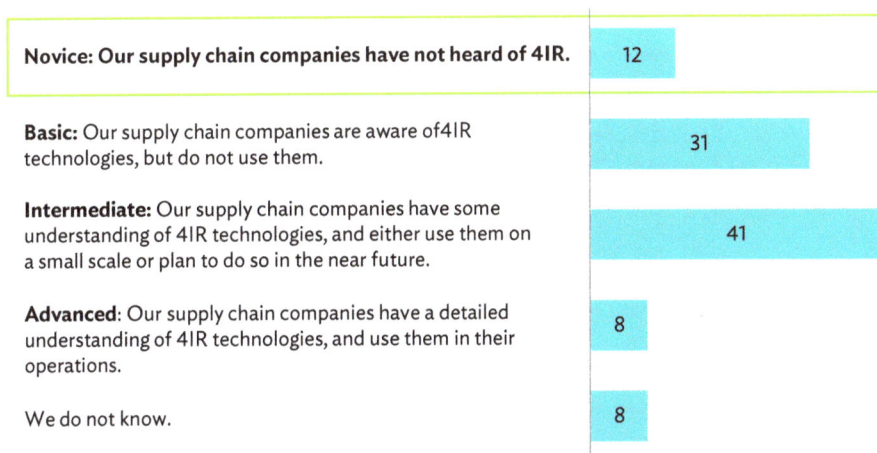

Novice: Our supply chain companies have not heard of 4IR.	12
Basic: Our supply chain companies are aware of 4IR technologies, but do not use them.	31
Intermediate: Our supply chain companies have some understanding of 4IR technologies, and either use them on a small scale or plan to do so in the near future.	41
Advanced: Our supply chain companies have a detailed understanding of 4IR technologies, and use them in their operations.	8
We do not know.	8

4IR = Fourth Industrial Revolution, IT-BPO = information technology-business process outsourcing.
Note: Based on survey of employers in the IT-BPO industry between June and September 2021 (n=51).
Source: Asian Development Bank Sustainable Development and Climate Change Department.

However, when compared against other economies with strong IT–BPO industries, the level of understanding of 4IR technologies in Punjab is relatively low. Past research estimates that 73% of IT–BPO firms in the Philippines have a good understanding of such technologies compared with only 53% in Punjab (Figure 16). Interviews with local experts and industry stakeholders found that the actual level of understanding and adoption of 4IR technologies by IT–BPO firms could be higher. The survey findings were likely influenced by firms in Pakistan associating 4IR technologies more closely with the manufacturing industry instead of the IT–BPO industry. When firms were probed on their use of specific technologies, it emerged that a large proportion of firms already experiment with or have adopted 4IR technologies. Only 4% of firms surveyed said they had not used cloud computing at all before and over 50% of firms intend to deploy cloud computing technologies across all possible functions by 2025 (Figure 19). This is likely due to most IT–BPO firms in Pakistan being focused on serving clients based outside of the country. The A.T. Kearney Global Services Location Index 2021 places Pakistan as one of the most financially attractive locations in the world for offshore services (AT Kearney 2021). Export remittances

Figure 18: Expected Increase in Output per Worker Due to Industry 4.0 Technologies between 2020 and 2025 in the Information Technology–Business Process Outsourcing Industry in Punjab, Pakistan

The majority of IT-BPO firms expect output per worker to increase by 50% to 100% in 5 years' time with the adoption of 4IR technologies

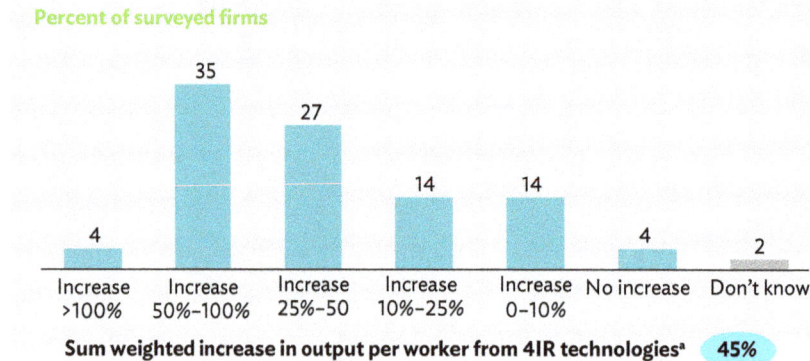

Percent of surveyed firms

Increase >100%	Increase 50%–100%	Increase 25%–50	Increase 10%–25%	Increase 0–10%	No increase	Don't know
4	35	27	14	14	4	2

Sum weighted increase in output per worker from 4IR technologies[a]   45%

Notes: Based on survey of employers in the IT–BPO industry between June and September 2021 (n=51). Calculated using sum-weighted average of output increase by the number of firms indicating different levels of expected increase in output, i.e., 0%, 0%–10%, 10%–25%, 25%–50%, 50%–100%, and over 100%. The midpoint of the range for each option for expected increase in output is used; for expected output increase of over 100%, the lower bound of 100% is used.

Source: Asian Development Bank Sustainable Development and Climate Change Department.

Figure 19: Current and Future Adoption of Relevant Industry 4.0 Technologies in the Information Technology–Business Process Outsourcing Industry in Punjab, Pakistan

More than 50% of IT-BPO companies plan to deploy cloud computing technologies across all possible functions in 5 years' time

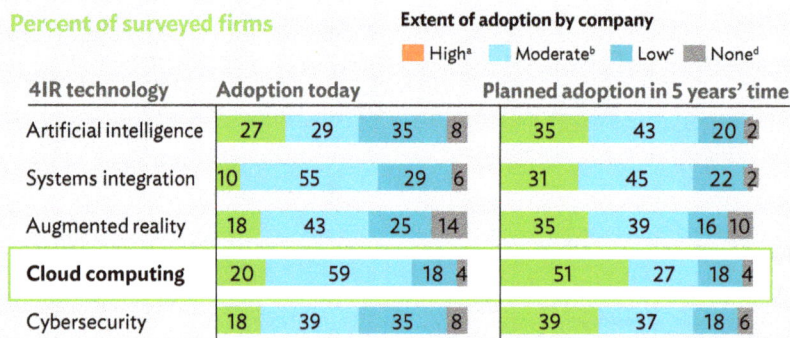

Percent of surveyed firms

Extent of adoption by company

■ High[a]   ■ Moderate[b]   ■ Low[c]   ■ None[d]

4IR technology	Adoption today				Planned adoption in 5 years' time			
Artificial intelligence	27	29	35	8	35	43	20	2
Systems integration	10	55	29	6	31	45	22	2
Augmented reality	18	43	25	14	35	39	16	10
Cloud computing	20	59	18	4	51	27	18	4
Cybersecurity	18	39	35	8	39	37	18	6

4IR = Fourth Industrial Revolution, IT–BPO = information technology–business process outsourcing.

[a] "High:" Firm has fully deployed the technology across all possible functions in the enterprise and/or has plans to fully deploy the technology across all possible functions in the future.

[b] "Moderate:" Firm has implemented the technology, but not fully deployed across all possible functions in the enterprise and/or plans to implement the technology across a few functions in the future.

[c] "Low:" Firm is experimenting with the technology at a very limited scale within the enterprise and/or plans to experiment with the technology in the future.

[d] "None:" Firm has not used technology at all within the enterprise and/or has no plans to use the technology in the future.

Note: Based on survey of employers in the IT-BPO industry between June and September 2021 (n=51).

Source: Asian Development Bank Sustainable Development and Climate Change Department.

from IT and IT-enabled services grew rapidly in 2020 (*The News* 2020). Research by BCG Platinion estimates that cloud computing technologies can cut the risk of data breaches by more than 95%, enabling Pakistan's IT–BPO firms to serve their foreign clients securely (*Google Cloud Blog* 2021).

Unsurprisingly, a large majority of IT–BPO firms believed that the COVID-19 pandemic will accelerate the use of 4IR technologies (Figure 20). Some firms attributed this to employees being incentivized to upskill through COVID-19-related retraining schemes. For instance, during the pandemic, the Punjab Skills Development Fund (PSDF) partnered online training provider Coursera to offer free online training courses to underprivileged youths in fields such as digital marketing, IT, and finance (PSDF 2020).

Figure 20: Perception on the Impact of the COVID-19 Pandemic on the Adoption of Industry 4.0 Technologies in the Information Technology–Business Process Outsourcing Industry in Punjab, Pakistan

The majority of IT-BPO employers believe that COVID-19 has accelerated or will accelerate the use of 4IR technologies

Percent of surveyed firms

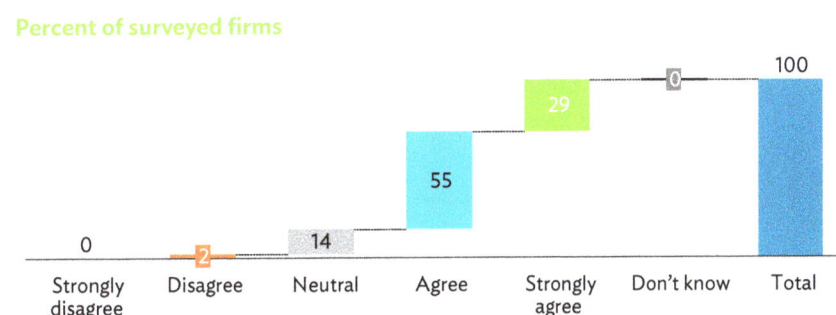

0	2	14	55	29	0	100
Strongly disagree	Disagree	Neutral	Agree	Strongly agree	Don't know	Total

Common reasons for accelerated adoption

Strategic shift toward greater digitization by company's management

Lack of labor due to movement restrictions necessitates more automation and shifting of activities to digital means

COVID=19 = coronavirus disease, IT–BPO = information technology–business process outsourcing.
Note: Based on survey of employers in the IT–BPO industry between June and September 2021 (n=51).
Source: Asian Development Bank Sustainable Development and Climate Change Department.

Skills Demand Analysis

Job Implications

The full adoption of 4IR technologies by firms in Punjab is expected to create 7,000 new jobs by 2025, or the equivalent of 18% of the 2020 IT-BPO workforce, which is over and beyond the BAU growth of the industry's labor force. This is due to the productivity effect again outweighing the displacement effect. Automation is expected to displace around 15,000 positions but create around 22,000 new jobs, over and beyond BAU growth of the industry's labor force (Figure 21). As in the textile and garment manufacturing industry, more IT–BPO jobs will be created in Punjab, Pakistan by 2025 even without 4IR. The IT–BPO workforce grew at approximately 20%

per annum[14] from 2015 and 2020. If this growth were extrapolated up to 2025, an additional 95,000 jobs could have been created even without adopting 4IR technologies. However, the full adoption of 4IR technologies could see a total of around 117,000 more IT–BPO jobs in 2025 than in 2020.

Figure 21: Estimated Impact of Industry 4.0 on Number of Jobs between 2020 and 2025 in the Information Technology–Business Process Outsourcing Industry in Punjab, Pakistan

The adoption of 4IR technologies will lead to 18% more jobs in 5 years' time in the IT-BPO sector

Percent of jobs impacted due to displacement and productivity effects of 4IR in 5 years' time

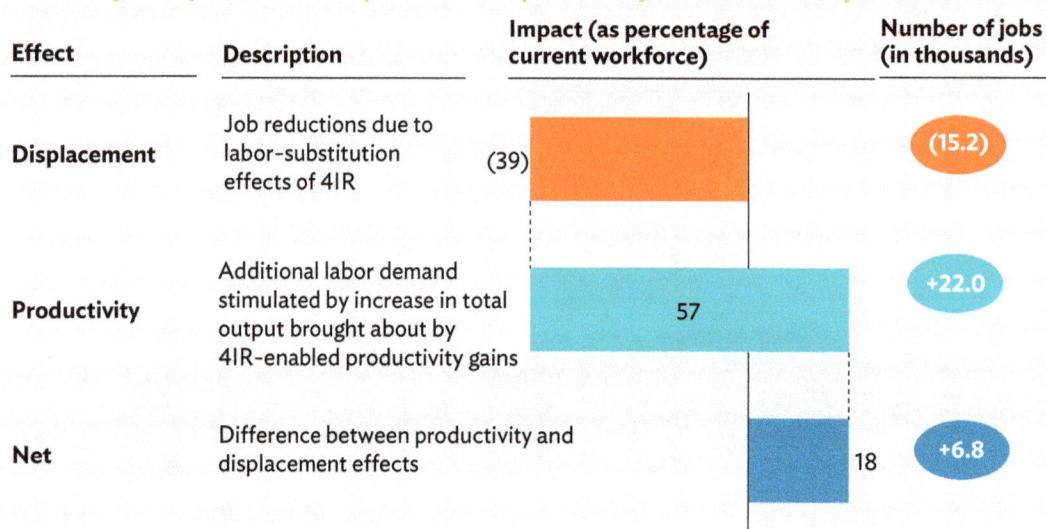

Effect	Description	Impact (as percentage of current workforce)	Number of jobs (in thousands)
Displacement	Job reductions due to labor-substitution effects of 4IR	(39)	(15.2)
Productivity	Additional labor demand stimulated by increase in total output brought about by 4IR-enabled productivity gains	57	+22.0
Net	Difference between productivity and displacement effects	18	+6.8

IT–BPO = information technology–business process outsourcing

Notes: Based on survey of employers in the IT–BPO industry between June and September 2021 (n=51). Industry employment data and gross domestic product output data are from Pakistan Bureau of Statistics. Numbers in parentheses ( ) means negative.

Sources: Asian Development Bank Sustainable Development and Climate Change Department and Pakistan Bureau of Statistics.

---

[14]   Calculated using industry and workforce data from Pakistan Bureau of Statistics and ILO. The information and communication sector was used as a proxy as there was no IT–BPO workforce data available.

Box 2: The Potential Cost of Government Inaction on Job Creation under Industry 4.0 Adoption

The net gains estimated in this report assume that the entire textile and garment manufacturing industry in Punjab, Pakistan adopts technologies of the Fourth Industrial Revolution (4IR or Industry 4.0). For this to occur, the government has a crucial role to play in ensuring that appropriate policies are in place to encourage participation, especially for firms that face significant adoption barriers. Effective policies are also needed to smooth labor market frictions, where public reskilling initiatives could be necessary for displaced workers to successfully transition into newly created jobs. An interesting analysis may be to understand how the breadth and effectiveness of policies impact net job creation under 4IR technology adoption. This will require empirical evaluations (e.g., simulations of impact on labor productivity growth under different policy scenarios) that are beyond the scope of this report and could be the subject of future research.

A simple thought exercise to provide some immediate guidance could be to assume that only the proportion of firms that indicated an intermediate or advanced understanding of 4IR technologies and their applications in 2020 (based on survey data from 2021), would reap the productivity gains of 4IR. In this case, only 35% of the 390,000 net job gains (around 136,500 jobs) expected in the full adoption scenario will be realized for the textile and garment manufacturing industry. Similarly, in the information technology–business process outsourcing industry, only 53% of the 6,800 jobs (around 3,600 jobs) expected to be created in the full adoption scenario will be realized.

This research further demonstrates that the impact of 4IR goes beyond job creation. 4IR will change the nature of jobs in the textile and garment manufacturing and IT–BPO industries and the types of skills that will be needed to take on these jobs. Past studies have demonstrated that this will free up time for workers doing routine tasks to take on higher-value tasks such as creative work. This would increase job satisfaction among workers and potentially lead to higher wages (AlphaBeta 2017).

Source: AlphaBeta. 2017. The Automation Advantage. https://alphabeta.com/wp-content/uploads/2017/08/The-Automation-Advantage.pdf.

As in the textile and garment manufacturing industry, the impact will differ by occupation. This study categorized jobs in the IT–BPO industry into five occupational groups (Table 5).

Table 5: Occupational Groups in the Information Technology–Business Process Outsourcing Industry

	Occupational Group	Examples of Job Titles
1	Technical	• Technician • Website designer
2	Managerial	• Chief executive officer • Human resources manager
3	Customer-facing	• Hotline operator • Customer service executive
4	Administrative	• Secretary • Finance executive
5	Elementary and/or manual jobs	• Data entry clerk • Cleaners

Source: AlphaBeta analysis.

Employers expect adopting 4IR technologies to bring the largest increase in technical occupations (Figure 22). The proportion of technical roles is expected to increase from 32.7% in 2020 to 35.8% in 2025 if 4IR technologies are fully adopted (Figure 23). This means that workers with the relevant technical skills will be sought after by employers.

As in the textile and garment manufacturing industry, most of the job gains created by the adoption of 4IR technologies will be gained by male workers (Figure 24). The differences are more pronounced in the IT–BPO sector. The number of jobs expected to be gained by male workers will exceed the number of jobs expected to be gained by female workers by 9.6 times. Women are not only less likely to benefit from the job gains due to 4IR, but the adoption of these technologies could add new challenges for female workers. Past research found that engineering and other STEM-related careers and subjects are associated with field work and long hours in Pakistan—both of which can be seen as unacceptable for a woman.[15] This means that female workers could find their career options further limited as the adoption of 4IR technologies increases the proportion of technical occupations. There would therefore need to be a strong focus on tech-related skilling programs for women in Punjab. For instance, in Indonesia, the Ministry of Tourism and Creative Economy launched "Coding Mum" in 2016, a two-month coding course to improve workforce participation rates of women (Aim2Flourish 2021). This scheme aims to attract mothers as well as female migrant workers returning from overseas back to the workforce by teaching them digital marketing and how to write basic programming codes that could allow them to set up their businesses digitally. Within three years of its launch, "Coding Mum" has trained over 600 graduates (Aim2Flourish 2021).

Figure 22: Expected Impact of Industry 4.0 on the Number of Jobs between 2020 and 2025 in the Information Technology–Business Process Outsourcing Industry in Punjab, Pakistan

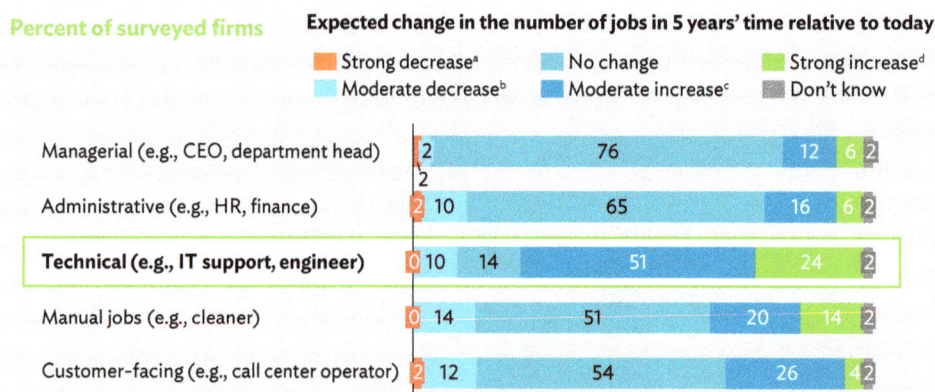

In the IT-BPO industry, technical jobs are expected to increase most significantly in 5 years' time due to the adoption of 4IR technologies

Percent of surveyed firms — Expected change in the number of jobs in 5 years' time relative to today

Strong decrease[a]   No change   Strong increase[d]
Moderate decrease[b]   Moderate increase[c]   Don't know

	Strong decrease	Moderate decrease	No change	Moderate increase	Strong increase	Don't know
Managerial (e.g., CEO, department head)	2	2	76	12	6	2
Administrative (e.g., HR, finance)	2	10	65	16	6	2
Technical (e.g., IT support, engineer)	0	10	14	51	24	2
Manual jobs (e.g., cleaner)	0	14	51	20	14	2
Customer-facing (e.g., call center operator)	2	12	54	26	4	2

CEO = chief executive officer, HR = human resources, IT = information technology.
a   Greater than or equal to 50% decrease in number of jobs.
b   Less than 50% decrease in number of jobs.
c   Less than 50% increase in number of jobs.
d   Greater than or equal to 50% increase in number of jobs.
Note: Based on survey of employers in the IT–BPO industry between June and September 2021 (n=51).
Source: Asian Development Bank Sustainable Development and Climate Change Department.

15   British Council. *Understanding Female Participation in Stem Subjects in Pakistan.* https://www.britishcouncil.pk/sites/default/files/stem_final_with_foreword_1_1.pdf.

Figure 23: Composition of Jobs in 2020 and by 2025 by Occupational Group in the Information Technology–Business Process Outsourcing Industry in Punjab, Pakistan

The distribution of jobs will change with technical jobs seeing the largest increase in 5 years' time

Weighted average percentage share of employees by occupational group in surveyed firms

○ Negative shift
○ Positive shift

Occupational group	Share today	Share in 5 years' time	Percentage shift
Administrative	15	14	–1.2%
Managerial	17	16	–1.2%
Customer-facing	17	16	–1.2%
Manual jobs	18	18	+0.5%
Technical	33	36	+3.1%

IT–BPO = information technology–business process outsourcing.

a   Average share of employees in surveyed firms is weighted by the number of employees in each firm, as indicated by respondents; percentages might not add up to 100% due to rounding.

b   The change in the number of workers in each job type is based on the number of firms indicating different levels of changes in number of jobs, i.e., "strong increase," "moderate increase," "no change," "moderate decrease," "strong decrease." The midpoint of the range for each option for expected change is used; for expected increase or decrease of over 50%, the low-bound of 50% was used.

Note: Based on survey of employers in the IT–BPO industry between June and September 2021 (n=51).

Source: Asian Development Bank Sustainable Development and Climate Change Department.

Figure 24: Estimated Net Job Gains by Gender from Industry 4.0 Adoption between 2020 and 2025 in the Information Technology–Business Process Outsourcing Industry in Punjab, Pakistan

Policies to increase female participation in the IT-BPO sector are critical to ensure that women also benefit from job gains from 4IR adoption

Estimated number of net jobs created by gender (in thousands)

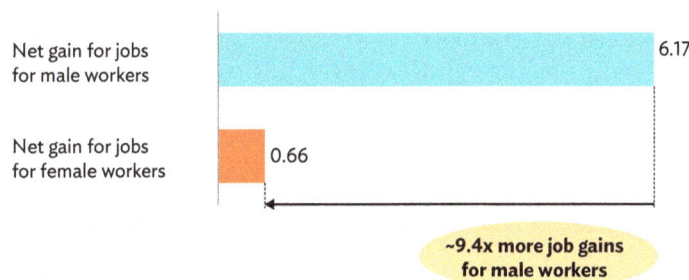

Net gain for jobs for male workers	6.17
Net gain for jobs for female workers	0.66

~9.4x more job gains for male workers

IT–BPO = information technology–business process outsourcing.

Notes: Based on survey of employers in the IT–BPO industry between June and September 2021 (n=51). Industry employment data and gross domestic product output data are from Pakistan Bureau of Statistics.

Sources: Asian Development Bank Sustainable Development and Climate Change Department and Pakistan Bureau of Statistics.

Task Implications

Similar to the textile and garment manufacturing industry, the analysis revealed only a slight drop in the proportion of time spent on routine tasks, both physical and interpersonal (Figure 25). Instead, more time will be allotted to analytical tasks which could involve creative work such as designing a new software or website, or designing customized cybersecurity solutions for clients. The limited change could be due to employers finding it difficult to visualize how the job of an average worker would change with 4IR adoption given that analytical work already takes up a fifth of weekly working hours in 2020 (based on 2021 survey data). Even without substantial shifts in the distribution of work, 4IR could lead to increased productivity and job satisfaction among IT–BPO workers. For instance, AI and data analytics technologies could better match workers' skills with tasks. Pakistani data and software company Afiniti introduced its advanced pairing solution to support the 5,000 call center agents at global telecommunications firm AT&T (*No Jitter* 2019). The AI-based software can match a customer with an agent with experience handling past customers with similar attributes, increasing productivity and customer satisfaction, and allowing customer agents to build up competencies in handling some customer groups.

Figure 25: Time Spent by Employees on Tasks at Work in 2020 and 2025 in the Information Technology–Business Process Outsourcing Industry in Punjab, Pakistan

Adoption of 4IR technologies may shift the distribution of weekly working hours toward analytical and nonroutine tasks

Average percentage share of weekly working hours spent by task in surveyed firms

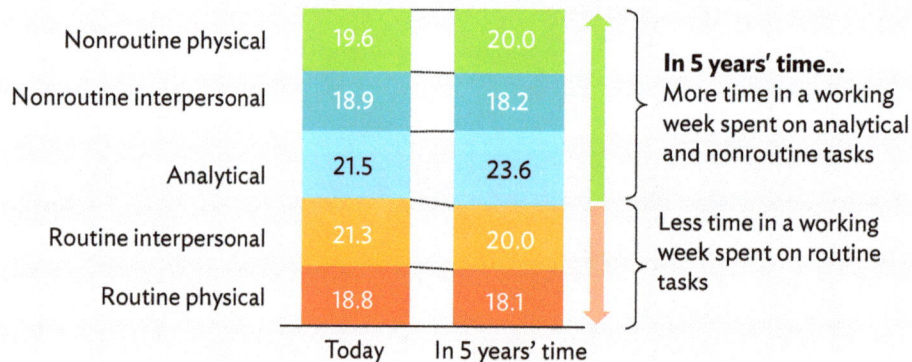

Task	Today	In 5 years' time
Nonroutine physical	19.6	20.0
Nonroutine interpersonal	18.9	18.2
Analytical	21.5	23.6
Routine interpersonal	21.3	20.0
Routine physical	18.8	18.1

In 5 years' time... More time in a working week spent on analytical and nonroutine tasks

Less time in a working week spent on routine tasks

IT–BPO = information technology–business process outsourcing.
Note: Based on survey of employers in the IT–BPO industry between June and September 2021 (n=51). Figures include rounding adjustments.
Source: Asian Development Bank Sustainable Development and Climate Change Department.

Skills Implications

The shift toward more analytical tasks is clearly reflected in the changes in the skills sought by employers by 2025. Creative thinking and design skills as well as complex problem-solving skills are the most valuable skills in 2025, jumping significantly in relative importance compared to 2020 (Figure 26). Social and interpersonal skills are expected to decline in relative importance, likely as 4IR technologies such as AI can replace human workers in some roles that require human interaction such as providing customer service support at call centers. The need for verbal communication is also reduced as digital platforms allow workers to collaborate remotely. Comparing the skills that are prioritized by employers for 4IR adoption by 2025 against skills in which a step-up in proficiency from a basic level of proficiency is needed, our estimates shows that workers would require a significant step-up in creative thinking and design skills from 2020 (Figure 27). Among the employers that consider a step-up from basic proficiency in creative thinking or design skills is needed, 49% indicated that a step-up to the intermediate level of proficiency is needed while 51% would like to see a step-up to advanced proficiency. A step-up to an advanced level of proficiency would be significantly more difficult to achieve and would require training institutions to ensure that trainers have the suitable capabilities and training. It could also mean that the education system would need to be pivoted to nurture creative thinking skills among students from an early age.

Figure 26: Importance of Skills in 2020 and for Industry 4.0 Adoption by 2025 in the Information Technology–Business Process Outsourcing Industry in Punjab, Pakistan

The relative importance of creative thinking, complex problem solving, and adaptive learning skills will increase very significantly in 5 years' time

Skills of increasing importance in 5 years' time
Skills of decreasing importance in 5 years' time
Skills with no change in importance in 5 years' time

Importance ranking	Today[a]	In 5 years' time[b]	Change in ranking
1	Verbal communication	Creative thinking or design	+7
2	Social and interpersonal	Complex problem solving	+5
3	Digital and/or ICT skills	Adaptive learning	+7
4	Critical thinking	Critical thinking	–
5	Written communication	Written communication	–
6	Management	Digital and/or ICT skills	– 3
7	Complex problem solving	Verbal communication	– 6
8	Creative thinking or design	Management	– 2
9	Numeracy	Numeracy	–
10	Adaptive learning	Social and interpersonal	-8

ICT = information and communication technology, IT–BPO = information technology–business process outsourcing.
[a] Evaluated using employer survey and supported by job portal data.
[b] Evaluated using the employer survey.
Notes: Based on survey of employers in the IT–BPO industry between June and September 2021 (n=51); and job data on the IT–BPO industry from the job portal Rozee. Figures include rounding adjustments.
Source: Asian Development Bank Sustainable Development and Climate Change Department.

Figure 27: Required Step-Up in Employee Proficiency Level for Industry 4.0 Adoption from 2020 to 2025 in the Information Technology–Business Process Outsourcing Industry in Punjab, Pakistan

To be 4IR-ready, workers would require proficiency leaps in numeracy skills

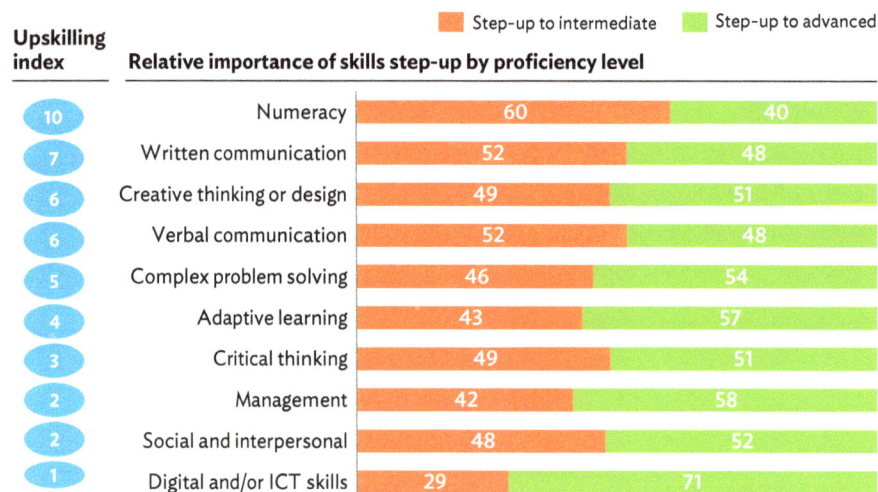

Upskilling index	Relative importance of skills step-up by proficiency level	Step-up to intermediate	Step-up to advanced
10	Numeracy	60	40
7	Written communication	52	48
6	Creative thinking or design	49	51
6	Verbal communication	52	48
5	Complex problem solving	46	54
4	Adaptive learning	43	57
3	Critical thinking	49	51
2	Management	42	58
2	Social and interpersonal	48	52
1	Digital and/or ICT skills	29	71

4IR = Fourth Industrial Revolution, BPO = business process outsourcing, ICT = information and communication technology.
Note: Based on survey of employers in the IT–BPO industry between June and September 2021 (n=51). Index is based on the number of employers indicating a need for workers with basic proficiency to be upskilled for each skill.
Source: Asian Development Bank Sustainable Development and Climate Change Department.

Skills Supply Trends

Like the textile and garment manufacturing industry, employers in the IT–BPO industry agree that there is a large variance in the quality of graduates available for hire but indicated that there were sufficient graduates to meet hiring needs (Figure 28). As in the textile and garment manufacturing industry, this is likely due to the demand for entry-level skills in the industry being at a relatively basic level, and the willingness of employers to provide training. During the country consultations and interviews conducted, industry stakeholders such as the Pakistan Software Houses Association for IT and ITeS (P@SHA) suggested that firms face more challenges in recruiting technically qualified candidates than reflected in the survey results. It was also highlighted that due to challenges in recruiting qualified, trained workers in the IT–BPO industry, firms would often offer traineeship programs to a large group of graduates and convert part of the cohort to permanent staff.

About 53% of firms strongly believe that their workers received an adequate amount of training (Figure 29). An average of 50% of workers receive on-the-job training across the industry in 2020 and this is expected to increase to 60% by 2025 (Figure 30). Interviews with local experts and stakeholders suggest that IT–BPO firms often provide at least basic training to workers as part of the induction process, to familiarize them with customized tools, platforms, or technologies. The P@SHA Skill Development and Training Committee partnered shortlisted training providers to offer training tracks on priority skill areas, such as programming languages, cloud technologies (e.g., AWS and Azure) and AI. Technical training and certification programs are offered to P@SHA member companies and their employees at subsidized rates.[16]

---

[16]   P@SHA. Inauguration of the P@SHA Skill Development Program 2021–2022. https://www.pasha.org.pk/pasha-skill-development-program-2021-22/.

Figure 28: Employer Sentiment Toward Graduates Hired in the Information Technology–Business Process Outsourcing Industry in Punjab, Pakistan

About 47% of employers strongly believe that there are sufficient graduates to meet their company's entry-level hiring needs

Percent of surveyed firms

Legend: Strongly agree | Neither agree nor disagree | Don't know or not applicable | Agree | Disagree or strongly disagree

Statement	Strongly agree	Agree	Neither agree nor disagree	Disagree or strongly disagree	Don't know or not applicable
There are sufficient graduates from relevant education/training programs to meet my company's entry-level hiring needs.	47	39	12	2	0
It is easy to identify and recruit high-quality graduates for entry-level positions at my company.	24	53	16	8	0
Graduates we hired in the past year were adequately prepared for the job by their education and/or training.	31	35	27	6	0
There is a large variance in the quality of graduates depending on region and education provider.	37	39	20	2	2
Graduates we hired in the past year have the appropriate "general" skills to be effective in entry-level positions, e.g., teamwork, creativity, problem-solving, etc.	27	43	24	6	0
Graduates we hire have the appropriate "job-specific" skills to be effective in entry-level positions, e.g., accounting skills, computer programming skills, etc.	31	49	16	4	0

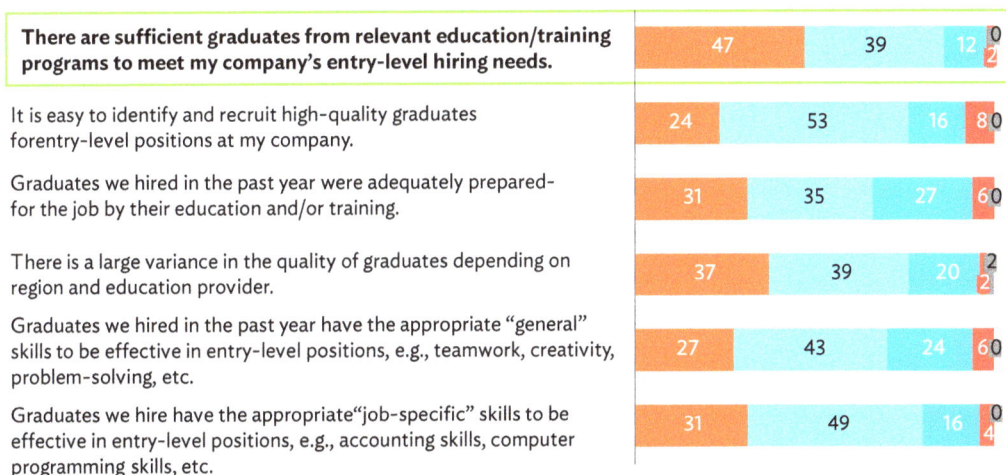

Note: Based on survey of employers in the IT–BPO industry between June and September 2021 (n=51).
Source: Asian Development Bank Sustainable Development and Climate Change Department.

Figure 29: Employers' Perception on Training for Employees in the Information Technology–Business Process Outsourcing Industry in Punjab, Pakistan

About 53% of employers strongly believe that their employees receive an adequate amount of training to do their jobs well

Percent of surveyed firms

Legend: Strongly agree | Disagree or strongly disagree | Agree | Don't know | Neutral

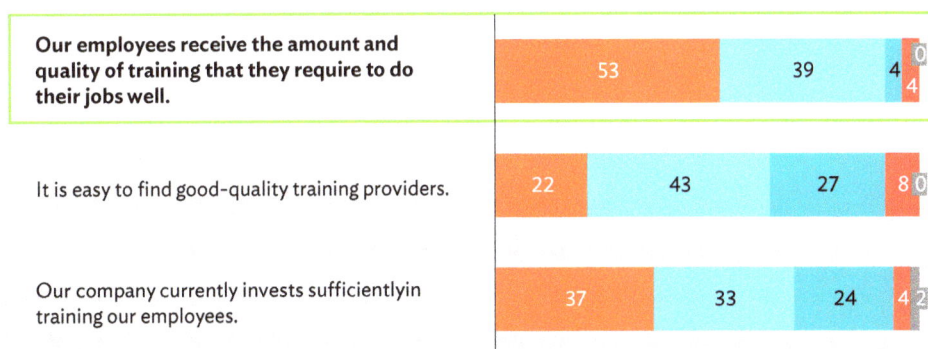

Statement	Strongly agree	Agree	Neutral	Disagree or strongly disagree	Don't know
Our employees receive the amount and quality of training that they require to do their jobs well.	53	39	4	4	0
It is easy to find good-quality training providers.	22	43	27	8	0
Our company currently invests sufficiently in training our employees.	37	33	24	4	2

IT–BPO = information technology–business process outsourcing.
Note: Based on survey of employers in the IT–BPO industry between June and September 2021 (n=51).
Source: Asian Development Bank Sustainable Development and Climate Change Department.

Figure 30: Proportion of Employees Receiving Training in 2020 and Requiring Training by 2025 for the Information Technology–Business Process Outsourcing Industry in Punjab, Pakistan

Half of IT-BPO workers receive on-the-job training today and employers estimate that 60% of workers will need such training in 5 years' time

Percentage share of employees by training channel

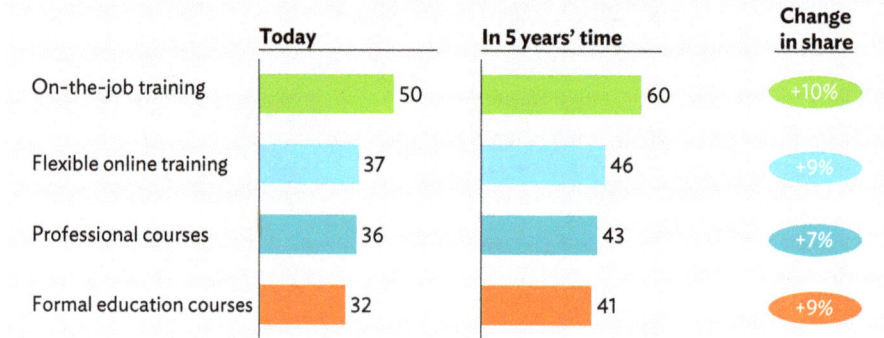

	Today	In 5 years' time	Change in share
On-the-job training	50	60	+10%
Flexible online training	37	46	+9%
Professional courses	36	43	+7%
Formal education courses	32	41	+9%

IT–BPO = information technology–business process outsourcing.
Notes: Based on survey of employers in the IT–BPO industry between June and September 2021 (n=51). The sum of all shares for 2020 and in 2025 time exceeds 100%, as one employee can undergo training in more than one training channel.
Source: Asian Development Bank Sustainable Development and Climate Change Department.

E. Emerging Jobs

Through the employer surveys conducted and scrapping of online job portals, it emerged that employers expect a variety of new job roles to become more prevalent as 4IR technologies are increasingly adopted across business functions. In the textile and garment manufacturing industry, jobs related to 3D printing and digital marketing are expected to be created (Figure 31). In the IT–BPO industry, these include big data specialists, machine learning experts, and cybersecurity engineers. Interviews with local experts and stakeholders in the IT–BPO industry also suggest that there is strong demand for workers with specific technical skills, such as Java developers, Ruby developers, and full-stack developers.

Figure 31: Job Roles Expected to Become More Prominent with the Adoption of Industry 4.0 Technologies between 2020 and 2025

The adoption of 4IR technologies will lead to different job roles in both industries according to employer surveys and online job portal analysis

Textile and garment manufacturing

3D printing specialist engineers: provide engineering support related to 3D printing and find new ways to use additive manufacturing (e.g., reducing the production time of garment design prototypes)

Digital marketing and/or e-commerce managers: engage clients through online channels to market textile products, design, and maintain website or other e-commerce channels

IT-BPO

Big Data specialists: set up infrastructure for data collection and analysis, integrate data from various resources, and manage big data on behalf of clients

Machine learning experts: develop artificial intelligence based algorithms and devices that enable machine learning (e.g., maintain AI-enabled chatbots)

Cybersecurity engineers: identify threats and vulnerabilities in systems and software, and develop and implement high-tech solutions to reduce the risk of breach on behalf of clients

4IR = Fourth Industrial Revolution, AI = artificial intelligence, IT-BPO = information technology-business process outsourcing.
Notes: Based on surveys of employers in the textile and garment manufacturing industry (n=52) and IT–BPO industry (n=51) between June and September 2021, and job data on the IT–BPO industry from the job portal Rozee (accessed 11 June 2021).
Source: Asian Development Bank Sustainable Development and Climate Change Department.

2 Overview of the Training Landscape

This chapter provides insights into the performance of the technical and vocational education and training (TVET) sector in Punjab, Pakistan as it prepares to deal with the challenges emerging from 4IR technology adoption. The insights are drawn from a survey of training institutions in Pakistan, focusing on Punjab, complemented with insights from the employer surveys discussed in Chapter 1.

Training institutions indicated a stronger understanding of 4IR technologies compared to employers, with 35% of training institutions surveyed strongly believing that they have a good understanding of skills required for 4IR. However, further probing showed that most training institutions in fact did not currently provide courses specific to 4IR technologies prioritized by the industry. For instance, over 50% of firms in both industries deploy AI technologies but only 21% of training institutions offer AI-related courses.

Correspondingly, there is scope to strengthen existing programs that encourage collaboration between employers and training providers to ensure that skills and knowledge taught in the classroom are relevant to industry needs. Currently, only 44% of training institutions gather inputs from industry stakeholders to design curricula, with employers in the IT–BPO industry participating more actively in curriculum design compared to textile and garment manufacturers.

To better understand the supply of talent and skills for the adoption of 4IR technology, a survey of 71 training institutions was undertaken in Punjab. These include public and private institutions of higher learning as well as TVET institutions. Institutions under the Technical Education and Vocational Training Authority (TEVTA) and Punjab Vocational Training Council (PVTC), two of the largest public training bodies in Punjab, were also included in the study. Of the training institutions surveyed, more than 90% trained at least 100 students per year. Around 45% provide courses relevant to the textile and garment manufacturing industry while over 60% offer courses relevant to the IT–BPO industry.

A. Industry 4.0 Readiness and the Impact of COVID-19

About 55% of institutions agree they have a good understanding of the skills that would be needed to be developed to prepare graduates for 4IR and more than half have plans to develop relevant training programs by 2025 (Figure 32). This is in comparison to only 35% of textile and garment manufacturers and 53% of IT–BPO employers having a good understanding of 4IR.

Figure 32: Perception of Training Institutions on Readiness for Industry 4.0 in Punjab, Pakistan

About 35% of training institutions strongly believe that they have a good understanding of the skills required for 4IR

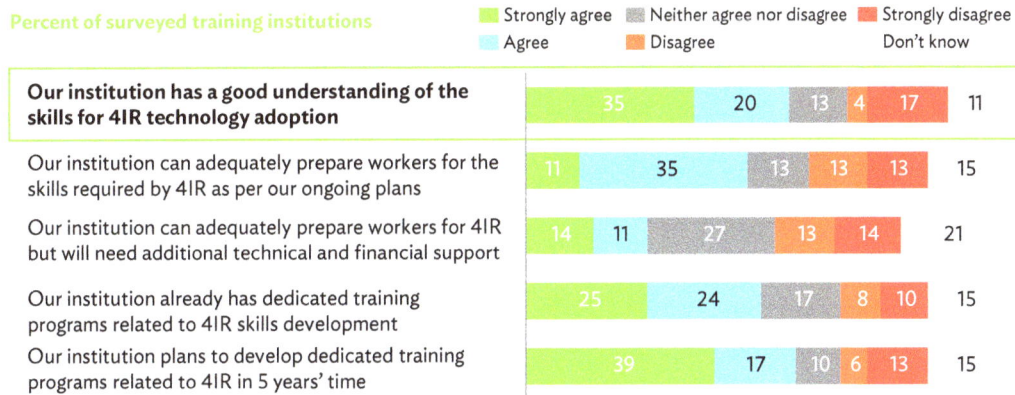

Percent of surveyed training institutions

Legend:
- Strongly agree
- Agree
- Neither agree nor disagree
- Disagree
- Strongly disagree
- Don't know

Statement	Strongly agree	Agree	Neither agree nor disagree	Disagree	Strongly disagree	Don't know
Our institution has a good understanding of the skills for 4IR technology adoption	35	20	13	4	17	11
Our institution can adequately prepare workers for the skills required by 4IR as per our ongoing plans	11	35	13	13	13	15
Our institution can adequately prepare workers for 4IR but will need additional technical and financial support	14	11	27	13	14	21
Our institution already has dedicated training programs related to 4IR skills development	25	24	17	8	10	15
Our institution plans to develop dedicated training programs related to 4IR in 5 years' time	39	17	10	6	13	15

4IR = Fourth Industrial Revolution.
Note: Based on survey of training institutions between June and September 2021 (n=71).
Source: Asian Development Bank Sustainable Development and Climate Change Department.

The movement restrictions imposed in Punjab and the rest of Pakistan during the COVID-19 pandemic affected most training institutions. In all, 75% of training institutions indicated that they had to close fully for some time due to inability to conduct in-person training (Figure 33). However, not all training institutions managed to pivot their activities and curricula to meet the changes in skills development needs created by the COVID-19 pandemic. Only 54% shifted courses online and only 20% adapted cost content to reflect new emerging skills needs. This is despite over half of textile and garment manufacturers and 84% of IT–BPO firms indicating that COVID-19 will accelerate the adoption of 4IR technologies.

Figure 33: Impact of COVID-19 on Training Institutions in Punjab, Pakistan

Only 20% of training institutions indicated that they altered course content to reflect new, emerging skills needs due to COVID-19

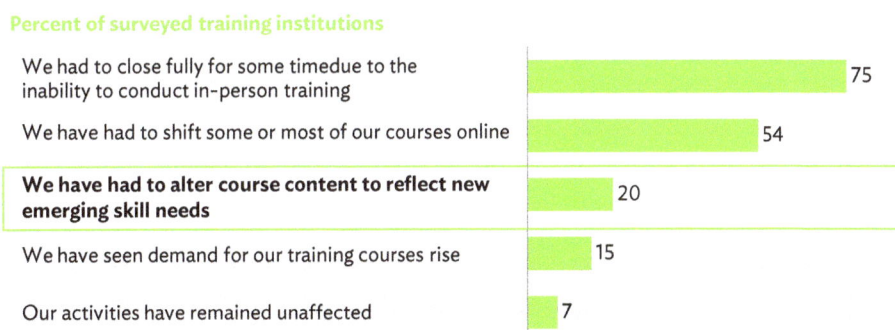

Percent of surveyed training institutions

Statement	Value
We had to close fully for some time due to the inability to conduct in-person training	75
We have had to shift some or most of our courses online	54
We have had to alter course content to reflect new emerging skill needs	20
We have seen demand for our training courses rise	15
Our activities have remained unaffected	7

COVID-19 = coronavirus disease.
Note: Based on survey of training institutions between June and September 2021 (n=71). Percentages do not add up to 100% as respondents were asked to select all options that apply.
Source: Asian Development Bank Sustainable Development and Climate Change Department.

B. Curricula

Industry 4.0 technologies can transform the workplace and jobs quickly. Hence, it is critical that training curricula be updated frequently to meet employers' needs. In Punjab, there is significant variance in the frequency at which training institutions updated their curricula. Figure 34 shows that 38% of training institutions update their curricula only once every 5–10 years or less frequently, while 34% update their curricula every 2–4 years. The remaining training institutions update their curricula every year or more frequently. This suggests that a small group of training institutions have a highly up-to-date curricula while most training institutions in Punjab might have difficulties responding to rapid changes in skills demand created by the shift to Industry 4.0.

Figure 34: Frequency of Review and Update of Curricula by Training Institutions in Punjab, Pakistan

Over a third of all training institutions review and update their curricula less frequently than once every 4 years

Percent of surveyed training institutions

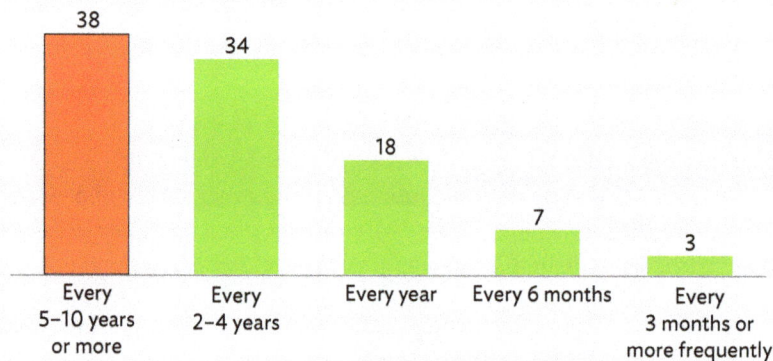

Note: Based on survey of training institutions between June and September 2021 (n=71).
Source: Asian Development Bank Sustainable Development and Climate Change Department.

The differences across training institutions are also demonstrated by the prevalence of 4IR related courses. Only 21% of the surveyed training institutions provide courses specific to 4IR technologies as of 2020 (Figure 35). Instead, training institutions in Punjab appear to be focused on building basic digital skills through broad-based programs to build general digital literacy (Figure 35). These gaps could pose challenges as more firms adopt 4IR technologies in specific applications. For instance, over 50% of firms in both industries deploy AI technologies but only 21% of training institutions currently offer AI-related courses (Figure 36). For these firms, courses focused on building general digital literacy are insufficient to produce workers able to operate AI-based systems. In addition, only a small proportion of training institutions use technologies such as online self-learning modules and augmented reality technologies or simulators to deliver training (Figure 35).

Interviews with local experts and stakeholders similarly found that there is scope for training institutions to offer more 4IR-related courses as well as use 4IR technologies more extensively to facilitate knowledge delivery. Increased exposure to industry-relevant technologies would help students to better adapt to the industrial environment. In addition, exposure to advanced 4IR technologies not yet adopted by firms in Pakistan could facilitate the introduction of new technologies into the industry when graduates eventually join the workforce. Targeted programs and policies to support training institutions to provide 4IR-relevant courses and 4IR-enabled training approaches could ensure that training institutions are able to cater to the future skill needs of employers.

Figure 35: Prevalence of Industry 4.0-Related Courses and Industry 4.0-Based Delivery in Training Institutions in Punjab, Pakistan

Currently, a relatively small share of training institutions teach specific to 4IR technologies and/or use 4IR technologies to deliver training

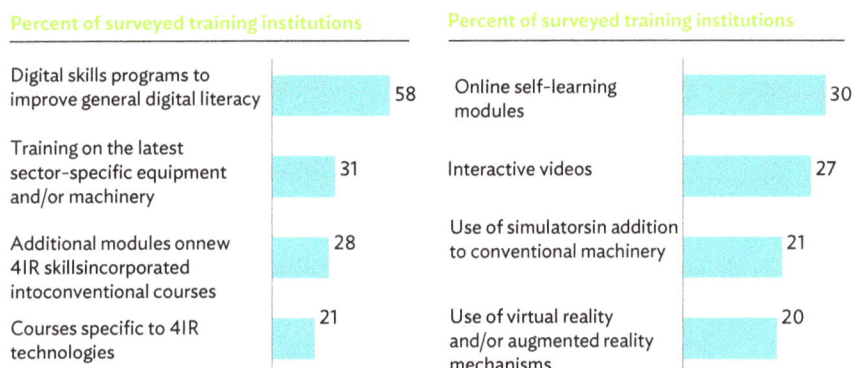

Percent of surveyed training institutions

Digital skills programs to improve general digital literacy	58
Training on the latest sector-specific equipment and/or machinery	31
Additional modules on new 4IR skills incorporated into conventional courses	28
Courses specific to 4IR technologies	21

Percent of surveyed training institutions

Online self-learning modules	30
Interactive videos	27
Use of simulators in addition to conventional machinery	21
Use of virtual reality and/or augmented reality mechanisms	20

4IR = Fourth Industrial Revolution.

Note: Based on survey of training institutions between June and September 2021 (n=71). Percentages do not add up to 100% as respondents were asked to select all options that apply.

Source: Asian Development Bank Sustainable Development and Climate Change Department.

Figure 36: Current Adoption of Specific Industry 4.0 Technologies by Employers and Prevalence of Courses Relevant to These Technologies in Training Institutions in Punjab, Pakistan

Over 50% of employers in both industries have deployed AI technologies but only 21% of training institutions offer courses in AI

Percent of employers that have deployed the technology;[a] Percent of surveyed training institutions[b]

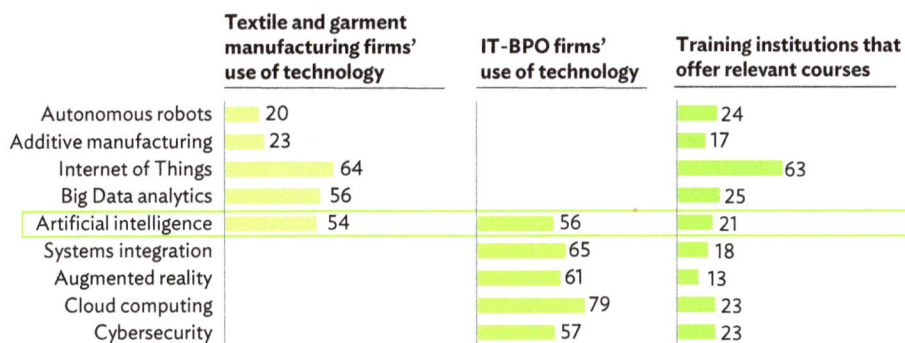

	Textile and garment manufacturing firms' use of technology	IT-BPO firms' use of technology	Training institutions that offer relevant courses
Autonomous robots	20		24
Additive manufacturing	23		17
Internet of Things	64		63
Big Data analytics	56		25
Artificial intelligence	54	56	21
Systems integration		65	18
Augmented reality		61	13
Cloud computing		79	23
Cybersecurity		57	23

AI = artificial intelligence, IT–BPO = information technology–business process outsourcing.

Notes: The percentage of employers that have deployed AI technologies is based on the percentage of firms that had responded "moderate" or "high" to current deployment of the technology. Percentages do not add up to 100% as respondents were asked to select all options that apply.

Sources: Based on surveys of employers in textile and garment manufacturing industry (n=52), IT-BPO industry (n=51), and training institutions (n=71) between June and September 2021; Asian Development Bank Sustainable Development and Climate Change Department.

C. Industry Engagement

Various policies and programs in Punjab encourage collaboration between industry stakeholders and training institutions. For example, the Punjab Technical Education and Vocational Training Authority (TEVTA), one of the province's public largest training providers, launched a textile and garment sector skills council to engage industry stakeholders in shaping training curricula and standards. However, only 37% of training institutions offer teaching placements for industry professionals and only 24% work with employers to provide industry placements for staff for training purposes (Figure 37).

There is scope for policies to encourage and facilitate more industry professionals to join the teaching profession, and more teachers or trainers to work directly with employers to ensure that students are taught by teaching staff with up-to-date and practical knowledge. Training institutions in Punjab show a strong focus on workplace-based training, with 73% of training institutions working with employers to organize workplace-based training for students, while over half organize industry apprenticeships (Figure 37). The new Apprenticeship Act aims to strengthen the relevance of apprenticeships in Punjab and global best practices can be adopted to ensure that apprenticeships prepare students for the workplace (*Dawn News* 2021).

Figure 37: Partnership Activities between Training Institutions and Employers in Punjab, Pakistan

Around three quarters of training institutions work with employers to organize workplace-based training for students

Percent of training institutions

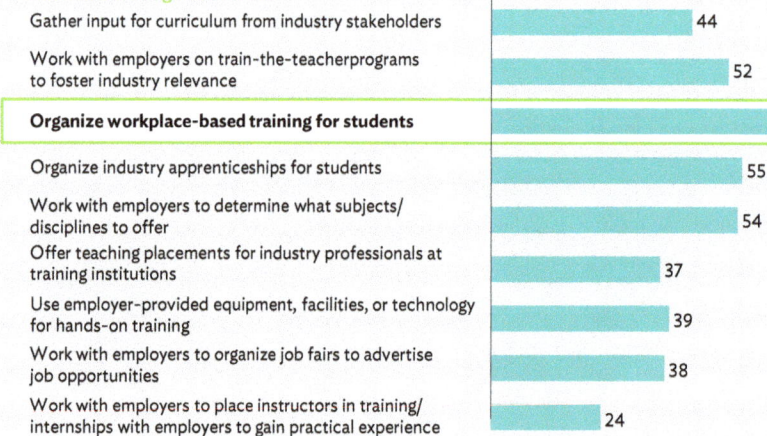

Activity	Value
Gather input for curriculum from industry stakeholders	44
Work with employers on train-the-teacher programs to foster industry relevance	52
Organize workplace-based training for students	**73**
Organize industry apprenticeships for students	55
Work with employers to determine what subjects/disciplines to offer	54
Offer teaching placements for industry professionals at training institutions	37
Use employer-provided equipment, facilities, or technology for hands-on training	39
Work with employers to organize job fairs to advertise job opportunities	38
Work with employers to place instructors in training/internships with employers to gain practical experience	24

Note: Based on surveys of employers in textile and garment manufacturing industry (n=52) and IT–BPO industry (n=51) between June and September 2021. Percentages do not add up to 100% as respondents were asked to select all options that apply.
Source: Asian Development Bank Sustainable Development and Climate Change Department.

The responses from training institutions on partnership activities contrasted with those from employers in the textile and garment manufacturing and IT–BPO industries. In both industries, apprenticeships similarly emerged as one of the key areas of cooperation, although less than half of employers surveyed engaged training institutions on workplace-based training (Figure 38). IT–BPO firms are more active in engaging training institutions compared to textile and garment manufacturers, with a higher number providing input on curricula and offering training opportunities to instructors. A higher proportion of IT–BPO employers also indicated that they engage employers several times a year or more frequently (Figure 39).

Figure 38: Partnership Activities between Employers and Training Institutions in Punjab, Pakistan

More than half of employers in both industries work with training institutions to offer industry apprenticeships

Percent of surveyed firms

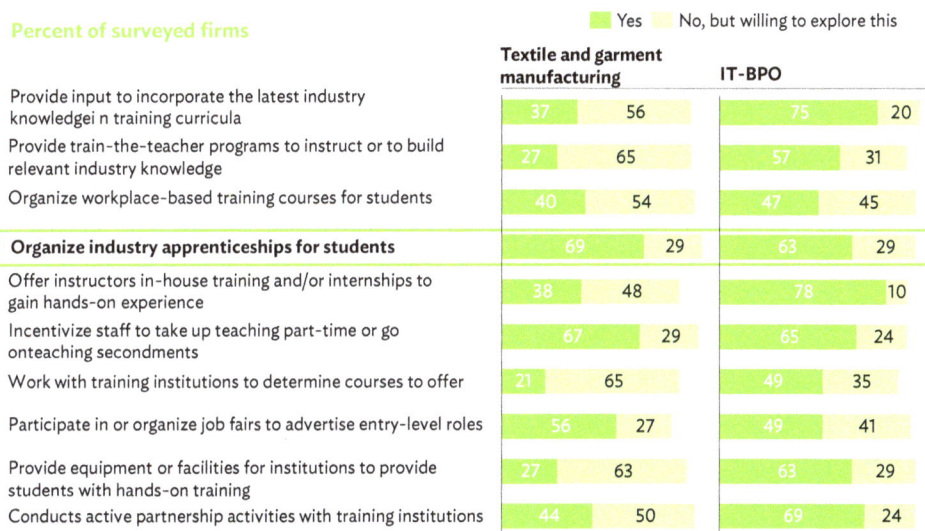

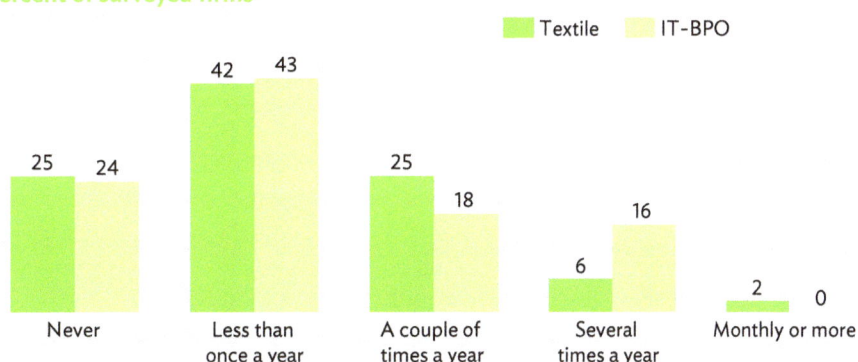

Legend: ■ Yes ■ No, but willing to explore this

	Textile and garment manufacturing		IT–BPO	
	Yes	No, but willing	Yes	No, but willing
Provide input to incorporate the latest industry knowledge in training curricula	37	56	75	20
Provide train-the-teacher programs to instruct or to build relevant industry knowledge	27	65	57	31
Organize workplace-based training courses for students	40	54	47	45
Organize industry apprenticeships for students	69	29	63	29
Offer instructors in-house training and/or internships to gain hands-on experience	38	48	78	10
Incentivize staff to take up teaching part-time or go on teaching secondments	67	29	65	24
Work with training institutions to determine courses to offer	21	65	49	35
Participate in or organize job fairs to advertise entry-level roles	56	27	49	41
Provide equipment or facilities for institutions to provide students with hands-on training	27	63	63	29
Conducts active partnership activities with training institutions	44	50	69	24

IT–BPO = information technology–business process outsourcing.

Note: Based on surveys of employers in textile and garment manufacturing industry (n=52) and IT–BPO industry (n=51) between June and September 2021.

Source: Asian Development Bank Sustainable Development and Climate Change Department.

Figure 39: Frequency of Communication between Employers and Training Institutions in Punjab, Pakistan

The majority of textile and garment manufacturing and IT-BPO employers communicate with training institutions less than once a year

Percent of surveyed firms

Legend: ■ Textile ■ IT–BPO

	Textile	IT–BPO
Never	25	24
Less than once a year	42	43
A couple of times a year	25	18
Several times a year	6	16
Monthly or more	2	0

IT–BPO = information technology–business process outsourcing.

Note: Based on surveys of employers in textile and garment manufacturing industry (n=52) and IT–BPO industry (n=51) between June and September 2021. Percentages do not add up to 100% as respondents were asked to select all options that apply.

Source: Asian Development Bank Sustainable Development and Climate Change Department.

D. Teachers, Trainers, and Instructors

Many training institutions reported themselves to be actively engaged in professional development and performance assessment of their teaching and training staff. Among training institutions, 89% reported that they provide professional development and training for their instructors. However, interviews with local experts and stakeholders in the IT–BPO industry suggest that school graduates lack relevant training to prepare them for 4IR, and further training might need to be provided for instructors to bridge this gap.[17]

Figure 40: Training Institutions' Practices to Support Instructors in Punjab, Pakistan

About 89% of training institutions provide professional development and training for instructors

Percent of surveyed training institutions

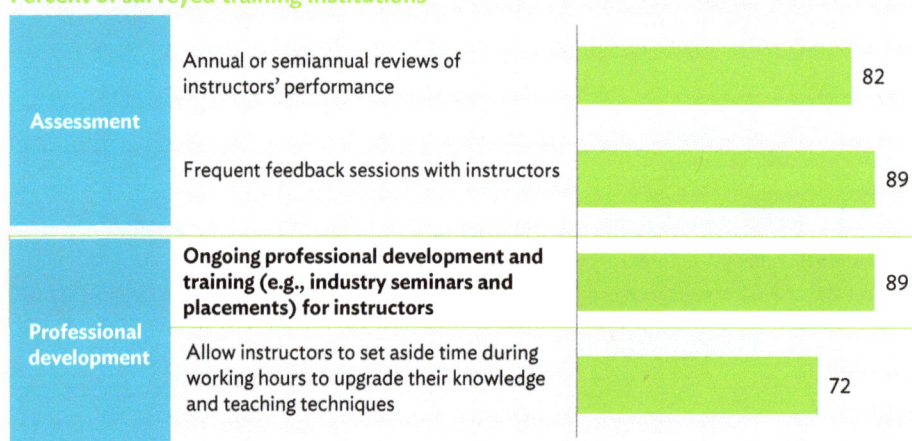

Assessment	Annual or semiannual reviews of instructors' performance	82
	Frequent feedback sessions with instructors	89
Professional development	Ongoing professional development and training (e.g., industry seminars and placements) for instructors	89
	Allow instructors to set aside time during working hours to upgrade their knowledge and teaching techniques	72

Note: Based on survey of training institutions between June and September 2021 (n=71). Percentages do not add up to 100% as respondents were asked to select all options that apply.
Source: Asian Development Bank Sustainable Development and Climate Change Department.

E. Performance and Policy Support

The training institutions were asked to comment on their current performance and the types of policy support that would be required. For 40% of the training institutions, it was somewhat difficult to fill training places, which they attribute to students believing that the skills gained from training would not be useful in their job search (Figure 41). In terms of impactful public policies, training institutions cited government funding for more students to take up courses and the building of quality assurance mechanisms as key policy priorities (Figure 42). During the interviews and country consultations conducted, local experts and stakeholders stressed that strong alignment on industry development and skills development policies across various government agencies was critical for the creation of quality jobs. For instance, to implement the e-Commerce Policy of Pakistan and create quality jobs in the e-commerce industry, the Ministry of Commerce worked closely with incubation centers and the Higher Education Commission to ensure that that training curricula is updated, and that relevant skills certifications are being put in place, so that graduates have the relevant skills to take on higher value-added roles in the industry (Commerce Division 2019).

---

[17]   Source: Country consultations conducted in September 2021.

F. Supply and Demand Mismatches

Figure 41: Training Institutions' Perceptions on and Reasons for Difficulty in Filling Places in Punjab, Pakistan

About 40% of training institutions find it difficult to fill places—the key reason is that students do not find skills taught to be relevant to their job search

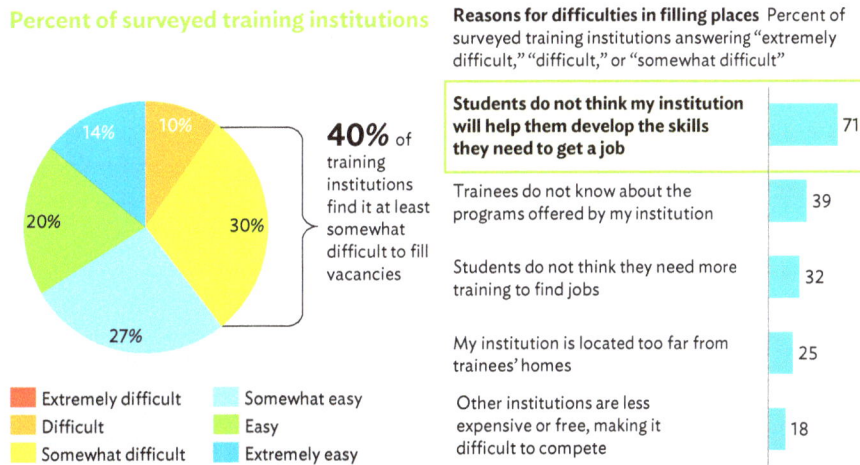

Percent of surveyed training institutions

Reasons for difficulties in filling places Percent of surveyed training institutions answering "extremely difficult," "difficult," or "somewhat difficult"

40% of training institutions find it at least somewhat difficult to fill vacancies

Pie chart values: 10%, 14%, 20%, 27%, 30%

Legend:
- Extremely difficult
- Difficult
- Somewhat difficult
- Somewhat easy
- Easy
- Extremely easy

Reason	Percent
Students do not think my institution will help them develop the skills they need to get a job	71
Trainees do not know about the programs offered by my institution	39
Students do not think they need more training to find jobs	32
My institution is located too far from trainees' homes	25
Other institutions are less expensive or free, making it difficult to compete	18

Note: Based on survey of training institutions between June and September 2021 (n=71). Percentages do not add up to 100% as respondents were asked to select all options that apply.
Source: Asian Development Bank Sustainable Development and Climate Change Department.

Figure 42: Training Institutions' Perceptions on Most Impactful Public Policies for Training Provision in Punjab, Pakistan

Policies related to greater flexibility in charging course fees and government funding for students are prioritized by training institutions

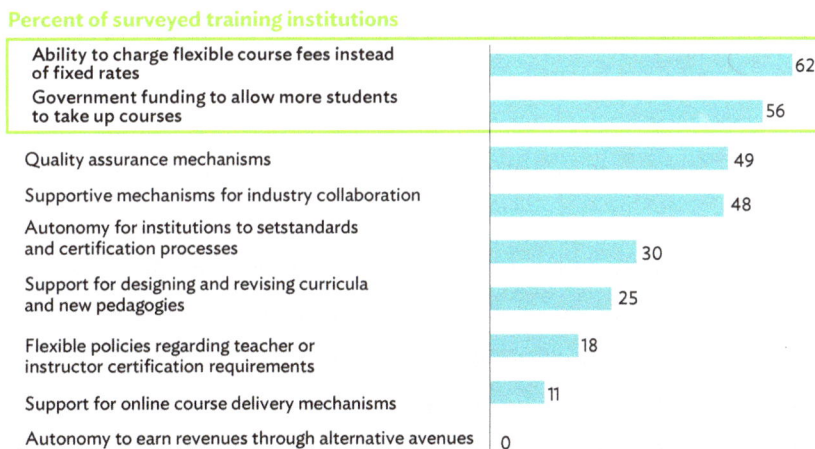

Percent of surveyed training institutions

Policy	Percent
Ability to charge flexible course fees instead of fixed rates	62
Government funding to allow more students to take up courses	56
Quality assurance mechanisms	49
Supportive mechanisms for industry collaboration	48
Autonomy for institutions to set standards and certification processes	30
Support for designing and revising curricula and new pedagogies	25
Flexible policies regarding teacher or instructor certification requirements	18
Support for online course delivery mechanisms	11
Autonomy to earn revenues through alternative avenues	0

Note: Based on survey of training institutions between June and September 2021 (n=71). Percentages do not add up to 100% as respondents were asked to select all options that apply.
Source: Asian Development Bank Sustainable Development and Climate Change Department.

Limited job opportunities and information asymmetry in the labor market were cited by training institutions as the most common reasons for their graduates being unable to find jobs (Figure 43). However, interviews with local experts and stakeholders suggest that job openings are available—it is the candidates in the workforce who lack the relevant skills (footnote 17). In particular, the IT–BPO industry faced challenges in filling job vacancies due to the lack of qualified candidates. This issue is likely to be exacerbated as the industry moves toward higher value-added services or knowledge process outsourcing functions. Greater proficiency in skills such as critical thinking would be needed (footnote 17). Thus, it is critical for training institutions to offer updated curricula that can meet the industry's needs. In addition, research indicates that employers in Pakistan rely on informal channels or personal networks to employ workers, and job portals are often outdated (Planning and Development Board 2015). Policies targeted at building a stronger awareness among students and training institutions of the jobs available and skills required could help to address any information asymmetry in the labor market.

There is also scope for training institutions to provide stronger support to graduates in their job search. Only 54% of training institutions surveyed provide information on job openings, and 62% provide support to students to prepare their job applications and attend interviews (Figure 44). This suggests that training programs are not focused on preparing workers for future employment in a large proportion of training institutions. In addition, only 34% of training institutions surveyed provided scholarships to students from disadvantaged backgrounds (Figure 44). This points to a need for policies focused on ensuring that socially disadvantaged communities have strong access to training opportunities.

Figure 43: Training Institutions' Perception of Reasons for Students Being Unable to Find Jobs upon Graduation in Punjab, Pakistan

Limited job opportunities and information asymmetry in the labor market are the key barriers to graduates' job search

Ranking based on responses from surveyed training institutions:

1 - Most common; 5 – Least common

Rank	
1	There are not enough job opportunities to incentivize job seekers to complete the relevant training to get a job.
2	There are not enough job opportunities.
3	There are enough jobs, but students do not know that employers have job openings.
4	Education and training programs do not adequately prepare job seekers for jobs.
5	The certifications provided to graduates of training institutions are not well recognized by employers.

Note: Based on survey of training institutions between June and September 2021 (n=71).
Source: Asian Development Bank Sustainable Development and Climate Change Department.

Figure 44: Non-Training Initiatives Provided by Training Institutions to Support Trainees in Their Professional and Personal Development in Punjab, Pakistan

More than half of training institutions work with employers to facilitate industry visits and exchanges

Percent of surveyed training institutions

Category	Initiative	Percent
Industry visits and exchanges	Visits from company representatives	66
	Visits to companies and potential employers	52
Career information and advice	Meetings with professional career coaches for career advice	62
	Provide information on job openings and salaries	54
	Provide statistics on program completion rates	44
	Provide alumni information (e.g., salaries, positions)	52
	Job application (e.g., CV preparation) and interview support	62
Financial and other support	Scholarships for students with disadvantaged backgrounds	34
	Meetings with counsellors for non-career advice (e.g., financial, personal)	56

Note: Based on survey of training institutions between June and September 2021 (n=71). Percentages do not add up to 100% as respondents were asked to select all options that apply.
Source: Asian Development Bank Sustainable Development and Climate Change Department.

The need for training institutions to strengthen their focus on employability is also reflected in the relatively poor perception that training institutions have of graduate quality. Only 61% of training institutions agree that their students have strong job-specific skills such as IT skills, accounting skills, or engineering skills required to take on specific roles (Figure 45). In contrast, 91% of textile and garment manufacturers and 80% of IT–BPO firms consider that the graduates they hired have such skills (Figure 45). This is in contrast to a similar study conducted in Southeast Asia, where on average, training institutions are much more optimistic about the preparedness of graduates for work than what employers report (ADB 2021). There could be a few reasons for this. First, as found through the employment surveys, the quality of graduates varies significantly across training institutions and employers might only hire the cream of the crop from a small proportion of institutions. Second, graduate quality in these industries could be generally higher as the demand for training in these industries exceed the demand for workers. Textile and garment manufacturing is one of the largest industries in Punjab while the IT-BPO industry is one of the fastest-growing. There is likely to be strong demand for training in these areas and employers can afford to select the best graduates. Insight from local experts and the Pakistan Software Houses Association (P@SHA) further suggest that notwithstanding the positive sentiments reflected in the surveys and the ability of some companies to select the best candidates, several IT–BPO firms face significant challenges in recruiting qualified candidates, particularly those with specific skills (e.g., full stack engineers). Third, employers are willing to provide on-the-job training, as seen from the analysis in Chapter 1. As of 2020, textile and garment manufacturers provide on-the-job training to 39% of workers and IT–BPO firms provide such training to 50% of workers. Therefore, these firms could have lower expectations of graduates' skills at entry level.

In addition, the positive perceptions of employers could change if training institutions are unable to pivot their programs to meet the new skills needs created by 4IR. The relative importance of various skills categories in both the textile and garment manufacturing and IT-BPO industries will change between 2020 and 2025 as discussed in Chapter 1. In the textile and garment manufacturing industry, digital and ICT skills will emerge as the most important skills while the relative importance of complex problem-solving skills will rise in the IT–BPO industry.

Figure 45: Perception of Employers on Graduates' Preparedness for Entry-Level Positions in Punjab, Pakistan

Compared to training institutions and IT-BPO firms, textile and garment manufacturers have a positive perception of graduates' preparedness.

Percent of surveyed training institutions and employers — Strongly agree — Agree

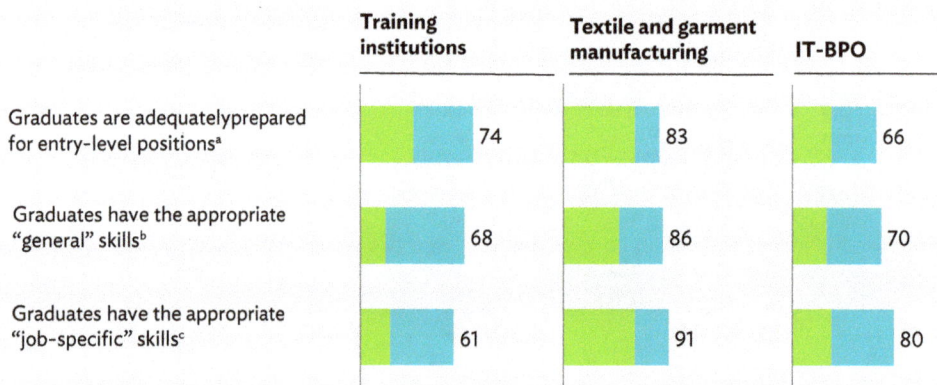

	Training institutions	Textile and garment manufacturing	IT-BPO
Graduates are adequately prepared for entry-level positions[a]	74	83	66
Graduates have the appropriate "general" skills[b]	68	86	70
Graduates have the appropriate "job-specific" skills[c]	61	91	80

[a] In their chosen fields of study.
[b] "General" skills include soft and generic skills that are developed through any academic program and experience and are requisite for success in any job, e.g., teamwork, creativity, and problem solving.
[c] "Job-specific" skills include skills relevant to the discipline or job description that are necessary to succeed in that specific position, e.g., accounting, computer programming, and engineering.

Notes: Based on surveys of employers in textile and garment manufacturing industry (n=52), IT–BPO industry (n=51), and training institutions (n=71) between June and September 2021.

Source: Asian Development Bank Sustainable Development and Climate Change Department.

3 National Policy Responses

A thorough scan of policies and programs across the government, industry, and civil society in Punjab, Pakistan reveals a range of efforts to advance Punjab into the technological era. A key government policy is the Punjab Growth Strategy 2023, which sets out strategic steps toward a knowledge-based economy, with technological innovation as a core pillar. There is also a strong focus on inclusive skills development—a key policy is the National "Skills for All" Strategy, which sets out reforms to strengthen TVET governance and funding and create better access to skilling for underserved groups. There are also various initiatives in Punjab aimed at upskilling the workforce digitally and encouraging a focus on skills over educational qualifications in labor markets. However, links between 4IR adoption and skill development are weak, and a single strategy encompassing both has not been formulated. 4IR trends and skills development needs are also not adequately coordinated across government, industry, and training providers. There is also scope for policy coverage to be broadened to include a stronger focus on building effective lifelong learning models and inclusive skilling policies to ensure that more workers benefit from Industry 4.0.

This policy assessment leverages government policy documents; consultations with government, industry, and civil society actors; academic literature; as well as the surveys and skills gap analysis conducted as part of this study.

A. Overview of Industry 4.0 Policy Landscape

The Punjab government has adopted a series of policies and road maps that set out Punjab's vision to build a vibrant ICT sector and skilled workforce, which will enable its shift toward inclusive knowledge-based economic growth. A key policy is the Punjab Growth Strategy 2023, which identifies the development of human capital as one of the key pillars for growth (Planning and Development Board 2019). Another important strategy is the Punjab Information Technology (IT) Policy 2018, which sets out a road map for the province's development into a regional ICT hub through initiatives including the promotion of entrepreneurship and improving digital literacy for vulnerable groups (Punjab Information Technology Board 2018).

The following list contains key 4IR and skills-related government policies in Punjab, including industry-specific policies relevant to the textile and garment manufacturing and IT–BPO industries. Table 6 summarizes the respective lead government entities for some of these policies.

(i) **Pakistan Vision 2025.** This is a development blueprint for Pakistan that includes developing human capital as one of its key pillars. Vision 2025 lays out plans to build the foundation of a knowledge-based economy that fosters innovation and entrepreneurship (Ministry of Planning, Development and Reform 2014).

(ii) **Digital Pakistan Policy.** The policy sets out Pakistan's plans to grow the ICT industry as well as to leverage ICT as an enabler in different sectors. It also sets out plans to accelerate the adoption of IOT, fintech, AI, and robotics technologies (Ministry of Information Technology and Telecommunication 2018).

(iii) **E-Commerce Policy of Pakistan.** The e-commerce policy is part of the Digital Pakistan Policy and provides guidelines for the promotion of e-commerce covering the regulatory environment, financial inclusion, ICT infrastructure, and youth and SME empowerment (Commerce Division 2019). It sets out plans for skills development programs covering areas such as digital marketing and capacity building and training for entrepreneurs, with a particular focus on female entrepreneurs.

(iv) **National "Skills for All" Strategy.** The strategy sets out recommendations for reforms and initiatives to strengthen TVET governance and funding, promote greater industry involvement in TVET, improve the quality and image of TVET, and create better access for underserved groups (Ministry of Federal Education and Professional Training 2018b).

(v) **Skills for Growth and Development: A Technical and Vocational Education and Training Policy for Pakistan.** This focuses on building an integrated TVET system with national standards for skills qualifications, assessment, and certification. It aims to design and deliver training programs to improve employability (Ministry of Federal Education and Professional Training 2018c).

(vi) **National Education Policy Framework 2018.** The policy includes plans to enhance TVET quality through private sector collaboration to strengthen the quality and relevance of skills training and improve certification and testing mechanisms. It also includes plans for the use of ICT in teacher training and classrooms (Ministry of Federal Education and Professional Training 2018a).

(vii) **Punjab Growth Strategy 2023.** The strategy sets out five key pillars that the Punjab government will focus on to drive growth, including (a) increased focus on the agriculture and SME sectors in which Punjab has comparative advantage, (b) private sector development, (c) development of human capital; (d) optimizing the allocation of public investment, and (e) coordination with the federal government on key policy interests. It sets out Punjab's vision of shifting toward knowledge-based economic growth through the development of the ICT sector and human capital (Planning and Development Board 2019).

(viii) **Punjab IT Policy 2018.** The policy sets out a road map for Punjab's development into a regional ICT hub. One of its goals is to strengthen the pool of ICT talent and improve general digital literacy among the population to increase employment opportunities (Punjab Information Technology Board 2018).

(ix) **Punjab Skills Development Sector Plan 2018.** The plan includes proposed initiatives to develop and expand skills training offerings, improve linkages between skills needs and training programs to increase employability of students, and improve access to training opportunities. It advocates strong collaboration with the private sector, including private training providers as well as employers in the implementation of the various initiatives (Planning and Development Board 2015).

(x) **RISE Punjab: Responsive Investment for Social Protection and Economic Stimulus.** This framework (Planning and Development Board 2020)sets out public investment strategies and priorities for Punjab post-COVID-19 that highlights the creation of a better trained and more productive workforce as an immediate priority. It sets out plans to pursue industry partnerships to provide demand-driven and forward-looking courses as well as establish centers of excellence.

(xi) **Textile and Apparel Policy 2020–2025.** The proposed policy aims to lift the textile and clothing exports significantly by 2025. Among other areas, it sets out plans to attract domestic and foreign investment in the textiles value chain and launch training programs for industrial stitching, focusing on women (Khan 2021).

Table 6: Key Policies Relevant to Managing the Impact of Industry 4.0 on Skills in Punjab, Pakistan

Policy Document	Responsible Entity	Relevance
Digital Pakistan Policy	Ministry of Information Technology and Telecommunication	Covers nationwide plans to leverage ICT as an enabler in different sectors such as agriculture, commerce, health, education, and public administration, as well as accelerate the adoption of IOT, fintech, AI, and robotics technologies
E-Commerce Policy of Pakistan	Ministry of Commerce	The policy sets out plans for skills development programs covering areas such as digital marketing and capacity building and training for entrepreneurs, with a particular focus on female entrepreneurs
National "Skills for All" Strategy	Ministry of Federal Education and Professional Training	Road map for skills development identifying areas where immediate interventions are required to improve the TVET system nationwide
Punjab Growth Strategy 2023	Planning and Development Board, Government of the Punjab	Punjab's growth strategy from 2018 to 2023, including planned interventions to improve the employability of Punjab's labor force through skills development and shift toward knowledge-based economic growth
Punjab IT Policy 2018	Punjab Information Technology Board	Road map for Punjab's development into a regional ICT hub, including plans to strengthen the pool of ICT talent in Punjab and improve general digital literacy among the population to increase employment opportunities
Punjab Skills Development Sector Plan 2018	Various government agencies in Punjab, including the Planning and Development Board; Department of Industries, Commerce and Investment; Punjab Technical and Vocational Training Authority; Punjab Vocational Training Council; and Punjab Skills Development Fund	Plan for skills development in Punjab including proposed initiatives to develop and expand skills training offerings, improve linkages between industry's skills needs and training programs, and improve access to training opportunities
RISE Punjab: Responsive Investment for Social Protection and Economic Stimulus	Planning and Development Board, Government of the Punjab	Framework guiding public investment strategies and priorities for Punjab post-COVID-19 that highlights the creation of a better trained and more productive workforce as an immediate priority

AI = artificial intelligence, COVID-19 = coronavirus disease, fintech = financial technology, ICT = information and communication technology, IOT = Internet of Things, IT = information technology, TVET = technical and vocational education and training.

Sources: Government of Pakistan, Ministry of Information Technology and Telecommunication. 2018. *Digital Pakistan Policy*. Islamabad. http://moib.gov.pk/Downloads/Policy/DIGITAL_PAKISTAN_POLICY(22-05-2018).pdf; Government of Pakistan, Commerce Division. 2019. *E-Commerce Policy of Pakistan*. Islamabad https://www.commerce.gov.pk/wp-content/uploads/2019/11/e-Commerce_Policy_of_Pakistan_Web.pdf; Government of Pakistan, Ministry of Federal Education and Professional Training. 2018. *National "Skills for All" Strategy*. Islamabad. https://www.giz.de/de/downloads/giz2019-en-tvet-national-skills.pdf; Government of Punjab, Planning and Development Board. 2019. *Punjab Growth Strategy 2023*. Lahore. https://pnd.punjab.gov.pk/system/files/PGS_2023%2019-21-145.pdf#overlay-context=economic_growth_strategy; Government of the Punjab, Punjab Information Technology Board. 2018. *Punjab IT Policy 2018*. Lahore. https://policy.pitb.gov.pk/system/files/Punjab_IT_Policy_2018_05062018.pdf; Government of Punjab, Planning and Development Board. 2015. *Punjab Skills Development Sector Plan 2018*. Lahore. https://pnd.punjab.gov.pk/system/files/Punjab_Skills_Sector_Plan_2018_0.pdf; Government of Punjab, Planning and Development Board. 2020. *RISE Punjab: Responsive Investment for Social Protection and Economic Stimulus*. Lahore. https://pnd.punjab.gov.pk/system/files/COVID-19%20Report%289jul2020%29_0.pdf.

B. Assessment of Current Policy Approaches in Punjab Related to Industry 4.0 and Skills

A diagnostic approach was taken to understand two important aspects of the 4IR skills policy approach in Punjab: (i) "the what"—the specific policies being adopted by Pakistan and the Punjab government and how they compare with international best practice approaches in preparing workers for 4IR; and (2) "the how"—the implementation mechanisms supporting 4IR efforts in government.

Assessment of Policy Actions ("The What")

Punjab's policies and programs were grouped into three action agendas that have been assessed to be most crucial to managing the impact of 4IR on jobs and skills.[18] Figure 46 shows the current degree of focus for each action area. The current degree of focus on each action area was rated as "strong," "moderate," or "weak," based on the analyzed extent of the policies' coverage in terms of scope and scale, as compared to those observed in international best practices.

Figure 46: Degree of Focus of Policy Actions to Manage the Impact of Industry 4.0 on Jobs and Skills in Punjab, Pakistan

Degree of focus of policy actions to manage the impact of 4IR on jobs and skills in Punjab, Pakistan

Degree of current focus: ■ Strong ■ Moderate ■ Weak

Action agenda	Key actions	Assessment
Stimulate 4IR adoption and worker reskilling efforts	Ensure strong and even adoption of 4IR across firms and workers	
	Build awareness of "in-demand" jobs and skills, as well as the benefits and opportunities of training	
	Incentivize employers and workers to participate in skills development	
	Foster close collaboration between governments, industry, and civil society to create relevant and effective nation-wide training frameworks	
Create new flexible qualification pathways	Establish effective lifelong learning models	
	Ensure relevance and agility of education and training curriculums to emerging skill needs	
	Encourage focus on skills rather than just qualifications in both recruitment and national labor market strategies	
Build inclusiveness to extend 4IR benefits to all workers	Build inclusive models that allow underserved groups to benefit from 4IR	
	Create social protection mechanisms for workers taking on flexible forms of labor	

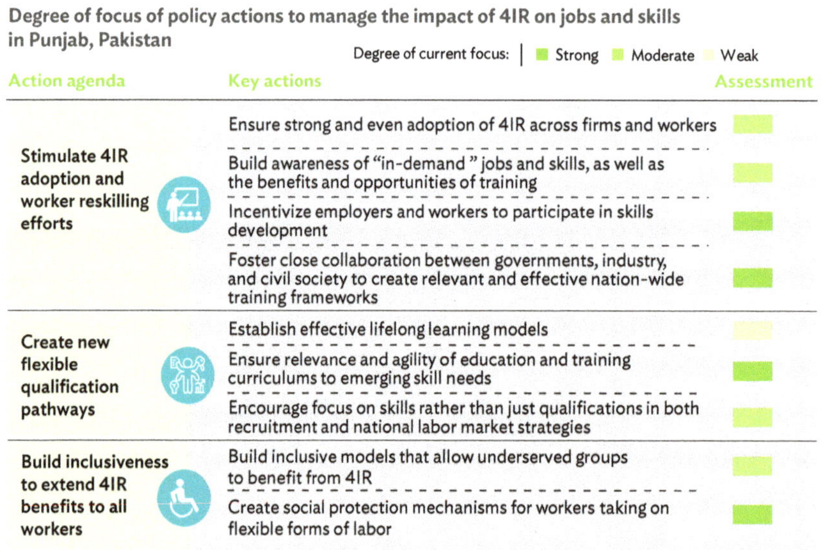

4IR = Fourth Industrial Revolution.

Notes: Degree of focus was assessed based on the following criteria: "Strong:" few or no gaps between the country's coverage of policy actions and coverage seen in international best practices; "Moderate:" medium level of gaps between the country's coverage of policy actions and coverage seen in international best practices; and "Weak:" significant gaps between the country's coverage of policy actions and coverage seen in international best practices.

Sources: Authors' representation; AlphaBeta analysis.

---

[18]   Based on AlphaBeta research on international best practices for policy actions that manage the impact of Industry 4.0 on jobs and skills. For details of these best practices, see Microsoft and AlphaBeta. 2019. *Preparing for AI: The Implications of Artificial Intelligence for Jobs and Skills in Asian Economies.* https://news.microsoft.com/apac/2019/08/26/preparing-for-ai-the-implications-of-artificial-intelligence-for-jobs-and-skills-in-asian-economies/.

Overall, the current degree of focus varies across the 4IR-relevant policy areas in Punjab. There is significant focus placed on policies to upskill workers digitally, as well as on initiatives that foster collaboration between stakeholders including government, industry, and civil society to ensure that the training curriculum is aligned with industry's practical skills needs. However, there is scope to strengthen policy focus on building effective lifelong models and collecting real-time data on jobs and skills demand and supply in the labor market. Specific efforts include the following:

(i) **Stimulating 4IR adoption and worker reskilling efforts.** Under the Hunarmand Jawan (Skillful Youth) program, the federal government plans to train 50,000 youth across Pakistan in technologies such as AI, cloud computing, and robotics (*Pakistan Today* 2020). The Pakistan Software Houses Association for IT and ITeS (P@SHA) works with training providers to offer technical training and certification programs in areas such as programming languages, cloud technologies (e.g., AWS and Azure), and AI to their members at subsidized rates.[19] In Punjab, there are similar efforts to build the workforce's digital capabilities. The Punjab Information Technology Board launched e-Rozgaar centers to provide training for youth in areas such as e-commerce and website development to enable them to work as digital freelancers.[20] The Punjab Skills Development Fund works with vocational training providers to fund courses and pay out stipends to incentivize trainees to complete the courses.[21] Training frameworks are created with industry inputs. TEVTA, one of the province's largest training providers, launched a textile and garment sector skills council to engage industry stakeholders in shaping training curriculums and standards (Malik 2020). However, there are some gaps that would need to be addressed. First, programs to build a stronger awareness of 4IR technologies and applications among firms, particularly in non-ICT sectors, are needed. The employer surveys show that only 35% surveyed firms in the textile and garment manufacturing industry have a good understanding of 4IR technologies and their applications. Second, real-time data on jobs and skills demand and supply in the labor market is needed. Many employers in Pakistan currently rely on informal channels or their personal networks to employ workers (Planning and Development Board 2015). This makes it difficult for graduates from training institutions to find employment due to a lack of widely accessible information about job openings and skill requirements.

(ii) **Creating new flexible qualification pathways.** Various programs are in place to align training curricula with industry needs, with a focus on skill-based qualifications instead of just educational qualifications. Under the Punjab Skills Development Fund (PSDF) Industry Training Program, businesses design training coursework and teaching materials based on their skills needs and can choose to conduct the course in their own facilities.[22] Similarly, under the PSDF's Skills for Garment program, training curricula is designed based on the input of garment manufacturers gathered through workshops, and employers are engaged directly as trainers to bring their best practices to the program.[23] TEVTA also works with employers to design training curricula focused on practical applications over theory (*The Nation* 2020). However, there are some gaps in ensuring that all workers have access to lifelong learning. The employer surveys show that the adoption of 4IR technologies is likely to increase the proportion of technical jobs in the industry and reduce the proportion of manual jobs. Manual, low-skilled workers at the highest risk of displacement would require reskilling to transit into new, technical roles created. However, 39% of the working population in Punjab is illiterate and may not be able to meet the educational qualification requirements of formal skills training providers to access reskilling opportunities (ADB 2020). There is a

---

19 P@SHA. Inauguration of the P@SHA Skill Development Program 2021-2022. https://www.pasha.org.pk/pasha-skill-development-program-2021-22/.

20 e-Rozgaar. About Us. https://www.erozgaar.pitb.gov.pk/#erb04.

21 Punjab Skills Development Fund. https://www.psdf.org.pk/

22 Punjab Skills Development Fund. Training Service Providers. https://www.psdf.org.pk/tsp/industry/

23 Study Solutions.Pk. PSDF Free Garments & Fashion Designing Courses With Monthly Stipend 2022. https://www.studysolutions.pk/free-garments-fashion-designing-courses-with-monthly-stipend.

need for stronger policy focus on programs that address the needs of illiterate or low-skill workers, who are at highest risk of being displaced by automation.

(iii) **Building inclusiveness to extend 4IR benefits to all workers.** Various programs have been implemented to digitally upskill women and disadvantaged youths. The Ehsaas Kafalat program provides low-income women with smartphones to facilitate digital financial transactions (*ARY News* 2020). The Ignite National Technology Fund administered by the Ministry of Information Technology and Telecommunication operates a network of incubators across Pakistan and runs the Digiskills Training Program, which aims to improve digital literacy of the population and empower uneducated youth and women.[24] The Consultative Committee on Women Economic Empowerment was formed by the Ministry of Commerce under the National E-commerce Council to explore initiatives to empower women through e-commerce and address the challenges faced by female entrepreneurs.[25] There are also policies to address the needs of freelancers. Pakistan is home to the world's fourth largest number of digital freelancers working on online platforms for contractual jobs (*Geo News* 2021, Qasim 2021). The Pakistan Software Export Board estimates that the exponential growth in the number of digital freelancers in Pakistan in recent years has resulted in a combined revenue of $150 million earned by freelancers in 2019. The Ministry of Information Technology and Telecommunication has drafted the National Freelancing Facilitation Policy 2021 that proposes fiscal incentives and subsidized health and life insurance to support freelancers and establish Pakistan as a leading global freelance market (Ministry of Information Technology and Telecommunication 2021). Despite efforts to ensure that the benefits of 4IR are extended to all workers, our analysis in Chapter 1 shows that workforce participation of women in the textile and garment manufacturing industry and IT–BPO industries in Punjab remains low. Women make up approximately half of Pakistan's population but were less than a quarter of the workforce in 2018.[26] There is scope to strengthen policies to ensure that women can also benefit from the job gains created by 4IR.

Assessment of the Implementation of 4IR Policies ("The How")

The implementation of a 4IR strategy for jobs and skills in Punjab, Pakistan was assessed against three dimensions found to be crucial for success according to past academic work: clarity and robustness of plans, strength of coordination between different stakeholders, and alignment of financing and incentives (Figure 47).[27]

Overall, the implementation approach varies across the critical dimensions. A 4IR road map that provides a framework for the adoption of 4IR technologies across all industries and sets out corresponding skills development needs has not been developed. Specific areas of concern are as follows:

(i) **Clarity and robustness of plans.** The Punjab Growth Strategy 2023 articulates Punjab's vision to build a knowledge-based economy through developing the ICT industry as well as encouraging technology adoption in other industries, including the textile and garment manufacturing industry (Planning and Development Board 2019). However, there is scope for policy makers in Punjab to devise clearer and more detailed policies on the transition to 4IR. First, there is currently no 4IR-focused vision or strategy in Punjab. The Punjab Growth Strategy 2023 mentions 4IR as one of the key trends that would impact industrialization but does not include an in-depth discussion on the impact of 4IR technologies and how these technologies could specifically change jobs and skill needs in Punjab. Second, there is limited

---

24   Digiskills. Who Should Join. https://digiskills.pk/WhoShouldJoin.aspx.
25   Source: Country consultations conducted in September 2021.
26   PBS. Labour force statistics. https://www.pbs.gov.pk/content/labour-force-statistics
27   Based on AlphaBeta research of Industry 4.0 strategies, plus insights from past public sector research, including Barber (2007) and McKinsey & Company (2012).

Figure 47: Implementation Challenges Associated with Industry 4.0 Policies for Jobs and Skills in Punjab, Pakistan

Implementation challenges associated with 4IR policies for jobs and skills in Punjab, Pakistan

Degree of current focus: | ■ Strong ■ Moderate □ Weak

Dimension	Questions	Assessment
Clarity and robustness of plans	Is there a clearly articulated vision for 4IR?	Weak
	Is there strong integration between employment or skills and the 4IR plan?	Weak
	Is the plan forward-looking, incorporating 4IR trends?	Weak
	Is there strong local data to support evidence-based policymaking?	
Strength of coordination	Is there one shared road map across industry and government departments for 4IR?	
	Is there coordination across different government ministries and levels?	Moderate
	Is there strong alignment within and between industry, and education and training institutions?	Strong
Alignment of financing and incentives	Is government financing aligned with the strategic goals?	Strong
	What is the strength of incentives for employers and workers to invest in skill development? What is the strength of incentives for teachers and institutions to ensure high-quality training and education systems?	Moderate

4IR = Fourth Industrial Revolution.
Notes: Degree of focus was assessed based on the following criteria: "Strong:" few or no gaps between the country's policy implementation approach and approach seen in international best practices; "Moderate:" medium level of gaps between the country's policy implementation approach and approach seen in international best practices; and "Weak:" significant gaps between the country's policy implementation approach and approach seen in international best practices.
Sources: Authors' representation; AlphaBeta analysis.

availability of strong local data to support evidence-based policy making.[28] The National Skills Information System was designed to provide accurate and up-to-date data on labor market needs of firms and the existing pool of skilled job seekers, but has had limited success in providing real-time data due to its lack of a robust organizational structure and poor institutional capacity (UNICEF 2019, ILO 2019b).

(ii) **Strength of coordination.** Overall, there are efforts to strengthen the links between skills development and employability in Pakistan. National and provincial sectoral skills councils incorporate employers' needs into training frameworks, and the PSDF supports partnerships between industry and training institutions through its Industry Training Programs.[29] About 73% of surveyed training institutions work with employers to organize workplace-based training for their students. However, interviews with local experts and stakeholders suggest that more coordination on 4IR-relevant policy areas is needed to strengthen Punjab's approach toward 4IR. First, a robust action framework that can capture the implications of 4IR for Punjab and Pakistan is missing.[30] This would need to be jointly developed by government agencies covering technology and innovation, education, and industry development at both the regional and national levels. Action plans for the adoption of 4IR technologies in key industries

---

[28] Sources include country consultations workshops conducted.
[29] *The News International*. 2018. Punjab Skills Development Fund. Industrial Training Programme. https://www.psdf.org.pk/industrial-training-programmes/; and Malik (2020).
[30] Sources include country consultations conducted in September 2021.

would need to be developed for Pakistan and Punjab to reap the potential gains of 4IR. Second, a single entity tasked with coordinating policies or plans related to 4IR adoption has not been identified.

(iii)   **Alignment of financing and incentives.** The budget allocation for education in Punjab falls within the recommended range of the United Nations Educational, Scientific and Cultural Organization (UNESCO) Education 2030 Framework for Action. In fiscal year 2021, Punjab allocated 17.6% of its public expenditure to education, within the range of 15% to 20% recommended by the UNESCO (*Daily Pakistan* 2020).[31] There have also been efforts to incentivize training institutions to improve the quality of training. In particular, the PSDF's Formal Training Programs that support training costs in registered institutions allow vocational institutions to access funds for training.[32] However, there is scope for stronger incentives to improve the quality of education. About 56% of training institutions surveyed would like to see more government funding for students to take up courses.

C. Assessment of Industry 4.0 Policies in Relation to the COVID-19 Pandemic

The COVID-19 pandemic has accelerated 4IR adoption in Punjab. For 51% of textile and garment manufacturing firms and 84% of IT–BPO firms, the COVID-19 pandemic has accelerated the use of 4IR technologies in their operations. With the use of 4IR technologies, the skills required by workers in these industries will have also shifted significantly, as the earlier chapters of this report have demonstrated.

The federal and Punjab governments have taken various steps to support the continuation of skills development during the COVID-19 pandemic in the immediate term; as well as prepare for the long-term skill shifts, with a focus on 4IR. During the COVID-19 pandemic, 54% of training institutions reported that they were able to shift courses online. These include TEVTA's institutions, which launched e-learning programs in areas such as AI, data analysis, and e-commerce training during the pandemic (Radio Pakistan 2020, PSDF 2020). The PSDF also worked with online training company Coursera to launch complimentary online courses during the pandemic. To prepare for the long-term skills shifts, the federal government partnered with the World Economic Forum (WEF) to launch the Parwaaz national accelerator program, under which the PSDF, as secretariat, is working with employers to launch skilling programs focused on emerging job roles (*Daily Times* 2021b). In Punjab, the RISE Punjab framework further lays out public investment strategies and priorities post-COVID-19. These priorities include creating a better trained and more productive workforce as an immediate priority and pursuing industry partnerships to provide demand-driven and forward-looking training courses (Planning and Development Board 2020).

Strategies in Punjab reflect a strong focus on the need to adapt and adjust economic growth plans in the post-COVID-19 environment. However, these policies might not be sufficient to address the needs of workers most vulnerable to the impacts of COVID-19 and increased automation. There is scope for the government to consider policies that cover the digital upskilling of workers to meet changing skills needs due to COVID-19. For instance, in Singapore, under the Redeployment / Job Redesign (JR) Reskilling program for rank-and-file workers, firms undergoing business transformation that will place their existing workers are at risk of redundancy or in vulnerable job roles can receive support to train their workers to take on new job roles or redesigned job roles within the same firm.[33]

---

[31]   *Daily Pakistan* (2020) and UNESCO. Education 2030: Incheon Declaration and Framework for Action for the implementation of Sustainable Development Goal 4. Paris. http://uis.unesco.org/sites/default/files/documents/education-2030-incheon-framework-for-action-implementation-of-sdg4-2016-en_2.pdf.

[32]   PSDF. Formal Training Programmes. https://www.psdf.org.pk/formal-training-programmes/.

[33]   Workforce Singapore. Rank-and-File Programmes Employers. https://www.wsg.gov.sg/rnf-place-and-train-programmes-employers.html.

4 The Way Forward

The previous three chapters highlight the potential benefits that 4IR can bring to Punjab, Pakistan as well as the challenges that would need to be addressed to achieve these benefits. This chapter identifies policy recommendations, based on global best practices, which policy makers in Punjab can consider adopting to unleash the potential opportunities created by Industry 4.0.

A. Summary of Key Challenges to Industry 4.0 Adoption Faced by Punjab, Pakistan

Table 7 provides a recap of the challenges as gleaned from the industry analysis (Chapter 1), the training institution survey (Chapter 2), and the policy assessment (Chapter 3).

Table 7: Recap of Challenges Facing Punjab, Pakistan in Relation to Industry 4.0

Area	Key Challenges	Findings
Textile and garment manufacturing industry	Limited understanding of 4IR technologies and their applications	Only **35%** of textile and garment manufacturers have a good understanding of 4IR technologies and applications
	Workers in manual roles are expected to face a higher risk of displacement while job gains are expected in technical roles	Over **70%** employers expect the proportion of technical roles to increase by 2025
	Job gains from 4IR will largely benefit male workers due to low female workforce participation	Only **one-third** of garment workers in Pakistan are female
IT–BPO industry	Workers would require a significant step-up in creative thinking and design skills	Creative thinking and complex problem-solving skills will become more important to IT–BPO employers
	Misalignment on skills demand with training institutions	**75%** of employers plan to adopt AR/VR technologies but only **13%** of training institutions offer courses in AR/VR technologies
	4IR could exacerbate the challenges faced by female workers in the IT–BPO industry as the number of technical jobs increase	STEM-related careers are associated with field work and long hours in Pakistan and seen as unsuitable for female workers

continued on next page

Table 7 *continued*

Area	Key Challenges	Findings
Training institutions	Limited use of 4IR technologies in the classroom	Only **20%** of training institutions use AR/VR technologies to deliver courses
	Graduates are unable to find jobs as information on jobs opening and employer skill needs are not available	Only **38%** of training institutions work with employers to organize job fairs
	Large variance in quality across different training institutions	Around **80%** of employers reported large variations in quality between graduates from different training institution
Policy assessment	Lack of clearly articulated 4IR vision integrated with jobs and skills	Plans to improve education and vocational training systems do not incorporate 4IR trends
	Need for stronger focus on skills instead of educational qualifications	**39%** of the working population in Punjab is illiterate and lack the basic educational qualifications required by training providers for continual reskilling
	Need for stronger targeted programs to ensure that women can reap the gains of 4IR in the long term	An estimated **11 million** women in Pakistan do not own a mobile phone due to family disapproval

AR/VR = augmented reality and/or virtual reality; 4IR = Fourth Industrial Revolution; IT–BPO = information technology–business process outsourcing; STEM = Science, Technology, Engineering, and Mathematics.
Source: Asian Development Bank Sustainable Development and Climate Change Department.

B. Recommendations to Address Challenges

To address the challenges faced by industry stakeholders in adopting 4IR technologies, and the gaps identified in the training landscape analysis and policy assessment, seven policy recommendations, drawing from international best practices, have been drawn up. These are summarized in Figure 48. Table 8 provides a summary of recommendations and potential lead agencies in Punjab to implement each recommendation as well as the approximate implementation time frame.

Table 8: Summary of Recommendations, Potential Lead Agencies, and Approximate Time Frame for Implementation

Recommendations	Potential Lead Agencies	Approximate Time Frame for Implementation
Develop 4IR adoption road maps for key sectors	Punjab Planning and Development Board	12–36 months
Develop innovative job-matching initiatives and platforms	Punjab Department of Labour, Punjab Information Technology Board	Less than 12 months
Strengthen existing frameworks for recognition of prior learning	Punjab Skills Development Authority, Trade Testing Board of Punjab	12–36 months
Promote adoption of 4IR technologies in knowledge delivery	Punjab Vocational Training Council, Punjab Technical Education and Vocational Training Authority	12–36 months

continued on next page

Table 8 *continued*

Recommendations	Potential Lead Agencies	Approximate Time Frame for Implementation
Strengthen relevance of industry apprenticeships and internships	Punjab Skills Development Authority	12–36 months
Inclusive 4IR reskilling policies	Punjab Planning and Development Board	12–36 months
Ensure responsiveness of education systems to changing skills needs	School Education Department, Punjab Planning and Development Board	12–36 months

4IR = Fourth Industrial Revolution.

Source: Asian Development Bank Sustainable Development and Climate Change Department.

Figure 48: Relevant Best Practices That Could be Adopted to Tackle Challenges in Adoption of Industry 4.0 Practices

4IR = Fourth Industrial Revolution; IT–BPO = information technology–business process outsourcing.

Source: Asian Development Bank Sustainable Development and Climate Change Department.

Recommendation 1: Develop Industry 4.0 Adoption Road Maps for Key Industries

Interviews with local experts and stakeholders suggest that greater coordination on 4IR-relevant policy areas is needed to strengthen Punjab's approach toward 4IR. Action plans for the adoption of 4IR technologies in key industries jointly developed by government agencies covering technology and innovation, education, and industry development, as well as industry stakeholders and training institutions could help to better align industry development and skills development plans.

The industry transformation maps (ITMs) developed by the Singapore government provide a useful model. Industry-specific ITMs have been developed for 23 industries, drawing together the inputs from private and public stakeholders in each industry, including trade associations and key firms. Each ITM charts out the overall growth direction for the industry under different pillars of transformation, such as jobs and skills, productivity, innovation, and internationalization. For example, the Aerospace ITM sets out Singapore's vision of building a globally recognized aerospace hub that is capable of design, engineering, production, and aftermarket services for the world's major aircraft programs.[34] The productivity pillar sets out the goals of adopting automation and IOT technologies in the industry, while the jobs and skills pillar sets out the objective of upskilling and reskilling existing and displaced workers. Policies and programs can take guidance from the ITM to help the industry achieve these objectives. Compared to existing plans in Punjab, Singapore's ITMs explicitly consider how innovation and technology will transform jobs and skills in each industry. In Punjab, separate strategies guide skills development, innovation, and the growth plans of specific industries. While the Punjab Growth Strategy collates these aspects into a single plan for Punjab's development toward a knowledge-based economy, it does not delve deeply into industry-specific priorities or the 4IR trends relevant to each industry. In Singapore's ITMs, specific industry trends and technological developments are considered and used to shape skilling plans.

Similar industry-specific road maps could be adopted in Punjab, coordinated by the Punjab Planning and Development Board. Policies related to technology adoption, industry development, and skills development often span a wide range of government stakeholders and a strong coordinating agency is needed to streamline these efforts and resolve any differences between stakeholders, as well as incorporate the needs of the private sector. A strong lead agency is therefore critical. For instance, Singapore's Aerospace ITM is led by the Economic Development Board, the lead agency for the sector, but the overall implementation is overseen by the Future Economy Council chaired by the deputy prime minister.[35]

Recommendation 2: Develop Innovative Job-Matching Initiatives and Platforms

A key challenge faced by policy makers and training institutions in Punjab is matching graduates to jobs. While the training surveys showed that graduates have difficulties finding jobs as information on job openings are not available, consultations with industry stakeholders found that employers in Punjab, particularly those in the IT–BPO industry, face significant challenges in finding qualified candidates. Innovative approaches to improve job-matching between employers and prospective workers could address this challenge (see Box 3).

The Punjab Department of Labour could consider working with technology agencies such as the Punjab Information Technology Board to launch platforms incorporating AI or Big Data technologies to strengthen job matching between workers and firms. Previous efforts to create online job portals have met with limited success with few job listings on portals created (*Daily Times* 2021a). As such, alongside creating job portals, policy makers would also need to embark on outreach efforts and work with industry association, business chambers, as well as training institutions to build awareness of the portals created.

---

[34]   Ministry of Trade and Industry. Aerospace. https://www.mti.gov.sg/ITMs/Manufacturing/Aerospace.
[35]   Ministry of Trade and Industry. ITMs Overview. https://www.mti.gov.sg/ITMs/Overview.

<hr>

Box 3: Developing Artificial Intelligence and Big Data-Powered Job-Matching Platforms in India, Malawi, and Singapore

In Singapore, an artificial intelligence jobs search portal has been launched to better connect jobseekers with career opportunities relevant to their skills. MyCareersFuture.sg seeks to reduce "missed matches" by recommending jobs that are best matched to the job title, skills, and minimum salary desired indicated by the user. The portal uses machine learning technology to scour various job descriptions posted in the jobs bank and to filter the skills required for each type of job. The users can then check off the skills they possess and view results of the search, where jobs that best correspond with the skills declared will show up at the top, each labeled with a percentage indicating the extent of match. For instance, a user who searches for the keyword "accountant" will be prompted with a list of relevant skills, such as "financial reporting," "budgets," "account reconciliation," and "Microsoft Excel," among others. The user will also receive recommendations of jobs in other sectors matched to the skills they had provided (*Today* 2018).

Artificial intelligence and Big Data are also used to obtain insights of the labor market in Malawi (European Training Institute 2021). Over 360,000 data points on myJobo.com, the largest jobsite in the country, were analyzed to provide key insights including the types of trending jobs and the specific skills sought by employers. Furthermore, such analysis supplements existing large-scale surveys like the Malawi Labor Force Survey and provides near-real-time information for jobseekers and training partners to keep up with labor market trends.

With the coronavirus disease (COVID-19) pandemic, recruitment fairs have also gone virtual. Platforms such as OnTime Job have built a model where recruiters can finish the whole hiring process on their mobile phone, giving them a choice to hire from anywhere and anytime. OnTime Job was launched in India in 2021 and conducted a large-scale virtual recruitment drive with over 8,000 participants and more than 50 successful job matches (*Higher Education Digest* 2020). Using data science and a mobile-first direct hiring approach, Industry 4.0 technologies can save time for recruiters and candidates, while driving better matches compared to traditional hiring processes (*Economic Times* 2020).

Sources: *Today*. 2018. New Jobs Portal Seeks to Reduce Job–Skills Mismatch. 17 April. https://www.todayonline.com/singapore/new-jobs-portal-seeks-reduce-job-skills-mismatch; European Training Institute. 2021. Big Data for Labour Market Intelligence. https://www.etf.europa.eu/sites/default/files/2021-06/guide_en_big_data_lmi_etf.pdf; *Higher Education Digest*. 2020. OnTime Job culminates India's Most Successful Live Recruitment Drive. 11 September. https://www.highereducationdigest.com/ontime-job-culminates-indias-most-successful-live-recruitment-drive/; *Economic Times*. 2020. AI-Powered Virtual Job Fairs Aim to Create Hassle Free Experience for Job Seekers and Employers. 15 September. https://hr.economictimes.indiatimes.com/news/workplace-4-0/recruitment/ai-powered-virtual-job-fairs-aim-to-create-hassle-free-experience-for-job-seekers-and-employers/78122272.

Recommendation 3: Strengthen Existing Frameworks for Recognition of Prior Learning

The lack of targeted policies to help low-skilled or illiterate workers upskill digitally is a key policy gap identified in Punjab. This gap is likely to be exacerbated with the adoption of 4IR technologies. Existing mechanisms for the recognition of prior learning would need to be strengthened and alternative pathways for education provided to ensure that low-skilled workers can continually reskill. The employer surveys show that in the textile and garment manufacturing industry, workers in manual occupations face the highest risk of displacement while jobs are most likely to be created in technical roles with the adoption of 4IR technologies. Displaced workers could face difficulties in reskilling themselves to take on new job roles as 39% of the working population in Punjab is illiterate and might not be able to meet educational qualification requirements of formal training providers (ADB 2020).

Currently, the Trade Testing Board of Punjab implements a recognition of prior learning (RPL) system to assess the skills gained by individuals through informal learning and grant them certification. However, while the evaluation process is focused on work experience, employer referrals and practical testing, educational qualifications are still considered.[36] Punjab could draw reference from the Malaysian Skills Certification Program to make the RPL system more inclusive. Under the program in Malaysia, skill certificates are granted to workers who do not have any formal educational qualifications but have obtained relevant knowledge, experience, and skills in the workplace to enhance their career prospects. Workers are only required to be able to speak and write in English and Bahasa Melayu, unlike under Punjab's existing system (see Box 4). The Punjab Skills Development Authority has been created to regulate skills development in Punjab and can take the lead to implement a robust RPL system.

Box 4: Malaysian Skills Certification Program

The "Malaysian Skills Certification Program" allows those who lack formal educational qualifications to enter their desired career paths. Recognized by industry, this program awards skills certificates at five different levels:

- Malaysian Skills Certificate (SKM) Level 1

- Malaysian Skills Certificate (SKM) Level 2

- Malaysian Skills Certificate (SKM) Level 3

- Diploma in Skills Malaysia (DKM) Level 4

- Malaysian Skills Advanced Diploma (DLKM) Level 5

These certificates are awarded across all sectors of the economy, which are classified into 22 sectors according to the country's National Occupational Skills Standard (OECD 2012). Importantly, no formal educational qualifications are required, according to the Department of Skills Department. Candidates may obtain these certificates through three channels: training in institutions accredited by the *Jabatan Pembangunan Kemahiran* (Department of Skills Development); industry apprenticeships under the "National Dual Training System"; and through sufficient "Accreditation of Prior Achievement" (see homepage of Department of Skills Department). The third channel refers to accreditation gained through evidence of past work and/or training experience.

With these certificates being accredited as officially recognized qualifications and mapped to equivalent academic qualifications under the Malaysian Qualifications Framework,[a] Malaysian companies can take guidance from this framework when assessing the suitability of job candidates without formal education but who possess the relevant skills to excel at the job (Malaysian Qualifications Agency 2019).

[a] The Malaysian Qualifications Framework consists of quality assurance standards that cover program design, objectives and learning outcomes, teaching, learning and assessment methodologies, support resources, and systems for delivery and improvement. Learning outcomes from programs submitted by higher education providers are verified and evaluated during periodic audit cycles (Malaysian Qualifications Agency 2019).

Sources: Malaysian Qualifications Agency. 2019. *Malaysian Qualifications Framework (MFQ). Second Edition*; OECD. 2012. *Skills Development Pathways in Asia*; and Department of Skills Department. 2022. *Malaysian Skills Certificate (SKM)*.

---

[36] Government of Punjab, Trade Testing Board. Recognition of Prior Learning. https://ttbp.edu.pk/recognition-of-prior-learning-rpl/.

Recommendation 4: Promote Adoption of 4IR Technologies in Knowledge Delivery

Industry 4.0 technologies can also help training institutions in Punjab improve their knowledge delivery. This is particularly important against the backdrop of changing movement restrictions and school closures forced by the COVID-19 pandemic. Currently, only a small proportion of training institutions in Punjab use interactive videos, self-learning online modules or VR technologies to deliver training.

Case studies in other countries demonstrate the potential of 4IR technologies in enhancing the delivery of virtual learning (see Box 5). Collaboration with the private sector could help to address the resource constraints faced by training institutions in Punjab in the adoption of 4IR technologies. Potential policy levers could

---

Box 5: Use of Artificial Intelligence, Virtual Reality, and Augmented Reality Technologies to Improve Learners' Experience in India, Germany, South Africa, and the United States

Industry 4.0 technologies can also be adopted by institutions to support students in their professional and personal development. For instance, a job interview simulator created by soft skills training specialist Bodyswaps teaches users various interview techniques and allows them to practice answering questions confidently, while using behavioral analytics to assess the user's verbal and nonverbal performance (Graham 2021). Such technologies could address problems such as a lack of trained staff at training institutions for dedicated career or interview coaching.

Virtual reality (VR) technologies also have specific applications to train students in some industries. In the information technology-business process outsourcing industry for instance, VR technologies can enable agents to experience realistic customer service scenarios that test and develop their capabilities, without exposing actual customers to agents who are not fully trained (Saeb 2017). In the industrial manufacturing space, German firm Siemens uses augmented reality to help trainees practice stimulated welding (Garage 2019, HR Technologist 2019).

Artificial intelligence (AI) technology can be used to develop customized training materials for each student based on their ability, preferred mode of learning, and pace of learning (Schmelzer 2019). In South Africa, the Department of Basic Education rolled out Ms Zora, an AI-based educational platform, to support the introduction of coding and robotics curriculum in schools. The AI-powered virtual assistant serves as both a teacher's assistant and personal tutor to pupils across all grades, enabling students to self-learn (iWeb 2020).

In the United States, schools are adopting augmented reality (AR) and other immersive technologies to build up soft skills such as creative problem solving and observation skills through different types of AR environments (zSpace, EdTechReview 2017, and Yourstory). In India, EdTech start-up fotonVR develops VR content for schools, providing a turnkey solution that includes the hardware, setting up the classroom, software, content, and training to the teachers. The cost of setting up a classroom, along with the content, kits, and training to teachers ranges from approximately $8,000–$20,000 (PRs1.8 million to PRs4.5 million) per classroom. Once the setup is done, fotonVR charges approximately $1,350–$2,700 (between PRs300,000 to PRs600,000) as an annual subscription charge.

Note: PRs = Pakistani Rupees

Sources: R. Schmelzer. 2019. AI Applications in Education. Forbes. 12 July; iWeb. 2020. Back to School: Robotics, Coding Curriculum Pushed Back. 15 January; zSpace. Beyond STEM: Building Soft Skills with Augmented and Virtual Reality; EdTechReview. 2017. 5 Schools That are Making the Most of AR & VR. 19 September; Yourstory. This Edtech Startup is Using Virtual Reality-Based Content to Have Real-Life Impact on Students; P. Saeb. 2017. Virtual Reality Potential for Training Contact Centre Agents. contact-centres.com. 3 January; R. Garage. 2019. Use Cases of Augmented Reality in Education and Training. 18 January; HR Technologist. 2019. How AR and VR are Revolutionizing Soft Skills Training in 2019. 22 April; P. Graham. 2021. Don't Stress About a Job Interview with Bodyswaps' New Simulator. BodySwaps. 2022. Job Interview Stimulator.

include pilot courses jointly developed by industry and training institutions. Box 6 also provides an example of "teaching factories" that have been implemented in Indonesia's TVET institutions. In Punjab, such pilots could be implemented by large public training providers such as the PVTC and the Punjab TEVTA to assess the suitability of specific tools or platforms in Punjab's context.

Box 6: Improving the Industry Relevance of Curricula in Indonesia's Technical and Vocational Education and Training institutions Through "Teaching Factories"

The "teaching factory" concept seeks to supplement theory-based education with practical demonstrations of production approaches in a simulated industrial facility (Lee Kuan Yew School of Public Policy and Microsoft 2016). Two models have been implemented in Indonesia:

- **Demonstration facilities in the institution that act as miniature replicas of actual factories where students learn to assemble and produce goods for industry partners.** With the support of the Asian Development Bank, a water treatment plant and bottle-filling machinery was set up in the State Polytechnic of Jakarta, which students learn to operate. Training institutions under Djarum Foundation's vocational education programs in Kudus, Central Java, operate school-based production units in which students, guided by teachers, produce goods and services that are sold to consumers and other businesses (Asia Philanthropy Circle and AlphaBeta 2017).

- **Utilization of the industry partner's actual factories to carry out demonstration lessons.** In collaboration with its industry partners PT Holcim Indonesia Tbk, PT Badak LNG, and LIGO Group of Companies, the State Polytechnic of Jakarta carries out lessons in these companies' factories where students observe and engage in actual factory operations for learning purposes.

Sources: Asia Philanthropy Circle and AlphaBeta. 2017. *Catalysing Productive Livelihood: A Guide to Education Interventions with an Accelerated Path to Scale and Impact.* http://www.edumap-indonesia.asiaphilanthropycircle.org/wp-content/uploads/2017/11/APC-Giving-Guide-Book-Final-Report-17112017.pdf; Lee Kuan Yew School of Public Policy and Microsoft. 2016. *Technical and Vocational Education and Training in Indonesia: Challenges and Opportunities for the Future.* https://lkyspp.nus.edu.sg/docs/default-source/case-studies/lkysppms_case_study__technical_and_vocational_education_and_training_in_indonesia.pdf?sfvrsn=e5c5960b_2.

Recommendation 5: Strengthen Relevance of Industry Apprenticeships and Internships

Industry apprenticeships and workplace-based training are prevalent in Punjab. Policies and frameworks to enhance the relevance of industry apprenticeship and workplace-based training could help make such programs more effectively. In Punjab, 73% of training institutions work with employers to organize workplace-based training for students and over half organize industry apprenticeships. Since 2020, the Higher Education Commission has made internships for university students mandatory, and all undergraduates are required to undergo a 9-week internship as part of course requirements (*Tribune* 2021). A new Apprenticeship Act that was recently passed in Punjab further aims to strengthen the relevance of apprenticeships in Punjab (*Dawn News* 2021).

Punjab could strengthen the relevance and effectiveness of internships and workplace-based training programs by establishing frameworks to guide the structuring and assessment of such programs (Box 7). The Punjab Skills Development Authority could work with industry associations to set up apprenticeship standards and assessment methods. This could include the All Pakistan Textile Mills Association for the textile and garment manufacturing industry, and the Pakistan Software Houses Association for IT and IT-enabled Services (ITES) for

---

**Box 7: Standards for Apprenticeships and On-the-Job-Training
in Singapore and the United Kingdom**

In Singapore, the Institute of Technical Education provides consultancy services to set up structured on-job-training programs according to the pedagogic competencies needs. Organizations that provide training to employees under approved such programs are eligible for government grants.

In the United Kingdom, apprenticeship standards are developed by an employer group, the Institute for Apprenticeships under the sponsorship of the Department of Education. The apprenticeship standard created in collaboration with employers sets out the skills, knowledge, and behaviors required of a qualified worker, while a separate document sets out how these are to be assessed at the end of the apprenticeship program. Curricula are then developed locally but must be consistent with the standards and the final assessment criteria (ILO 2020a). For instance, the standards for garment makers set out specific duties that apprentices must be able to undertake (e.g., quality check garment components and materials before, during, and after completion of the garment); and methods of assessment (e.g., garment components must meet quality standards throughout production). It also sets out the specific skills, Institute of Technical Education, as the ability to assemble fabric components to make a whole garment, and behavior, such as taking a health and safety-first attitude, required by apprentices.

Sources: International Labour Organization (ILO). 2020. *ILO Toolkit for Quality Apprenticeships*. Geneva. https://www.ilo.org/wcmsp5/groups/public/---ed_emp/---ifp_skills/documents/publication/wcms_751114.pdf; ITE Singapore. Industry Training Schemes. https://www.ite.edu.sg/employers/industry-training-schemes/certified-on-the-job-training-centre.

---

the IT–BPO industry. Public sector training providers such as TEVTA and PVTC could take the lead in working with firms to set up strong internship or in-house on-job training programs that incorporate best practices in training administration.

Recommendation 6: Adopt Inclusive Industry 4.0 Reskilling Policies

Women make up approximately half of Pakistan's population but were less than a quarter of the workforce in 2018. Women are not only less likely to benefit from the job gains due to 4IR, but the adoption of these technologies could add new challenges for female workforce participants as the proportion of technical jobs in industries increases. Technical jobs such as engineering and other STEM-related careers are associated with field work and long hours in Pakistan—both of which can be seen as unsuitable for a woman. Past research further shows that women in Pakistan face unique challenges in accessing skilling and employment opportunities. The distance of training facilities and availability of secure transport have also heavily impacted the ability of women to access training (Planning and Development Board 2015). Social norms do not support women's involvement in economic activities and employment options are largely limited to roles in education and health care, or fields that do not require regular interaction with nonfamily men (Planning and Development Board 2015). Policy makers in Punjab can consider initiatives to build stronger digital literacy among women; and women to empower themselves through technology (see Box 8). This could be done in collaboration with international partners, such as ILO or ADB. These initiatives could be coordinated by the Punjab Planning and Development Board.

Box 8: Providing Reskilling Opportunities for Women in Cambodia, India, and Indonesia

In Cambodia, the United Nations Educational, Scientific and Cultural Organization partnered with the Garment Manufacturers Association in Cambodia to support lifelong learning among garment factory workers across Cambodia under the Factory Literacy Program (*The Phnom Penh Post* 2021). Under this program, female factory workers are equipped with functional literacy and numeracy skills. Those who pass the course's final exam obtain a certificate of completion accredited by the Ministry of Education, Youth and Sport.

In India, the Disha project, a partnership between the United Nations Development Programme, IKEA Foundation, and the India Development Foundation creates employment and entrepreneurship opportunities for women (Deloitte 2019). Disha supports various programs to provide career counselling, skills training, and job placements for women. For instance, women in Telangana were trained to use computer-aided design software to design sarees (traditional women's clothing in India) and then sell them to master weavers. They are also trained on entrepreneurial skills and to support other women in starting their own businesses. Disha was able to emerge as an impactful initiative on account of its strong partnerships with government, businesses, training institutions, and civil society (Deloitte 2019).

In Indonesia, the Ministry of Tourism and Creative Economy launched "Coding Mum" in 2016, a two-month coding course to improve workforce participation rates of women (Aim2Flourish 2021). This scheme aims to attract mothers as well as female migrant workers returning from overseas back to the workforce by teaching them digital marketing and how to write basic programming codes that could allow them to set up their businesses digitally. Within 3 years of its launch, "Coding Mum" had trained over 600 graduates (Aim2Fluorish 2021).

Sources: Deloitte. 2019. *Empowering Women and Girls in India for the Fourth Industrial Revolution.* https://www2.deloitte.com/content/dam/Deloitte/my/Documents/risk/my-risk-sdg5-empowering-women-and-girls-in-india-for-the-fourth-industrial-revolution.pdf; Aim2Flourish. 2021. *The True Definition of Wonder Woman.* https://aim2flourish.com/innovations/the-true-definition-of-wonder-woman; *The Phnom Penh Post.* 2021. Factory Literacy Programme Expands. 29 March. https://www.phnompenhpost.com/national/factory-literacy-programme-expands.

Recommendation 7: Ensure Responsiveness of Education Systems to Changing Skills Needs

This research revealed that the skills sought after by employers will change significantly with the adoption of 4IR technologies. Creative thinking and/or design skills, and complex problem-solving skills will become the most sought-after skills for IT–BPO employers by 2025 while digital and/or ICT skills will become most valued by employers in the textile and garment manufacturing industry. Soft skills such as complex problem-solving and critical thinking will become more valued by employers than traditional skillsets such as written communication. Against the backdrop of 4IR, there is strong imperative to ensure that education systems, including at the primary and secondary levels, are constantly updated to respond to changing skill needs.

Past studies have pointed to a poor quality of education for students in Punjab even at the foundational levels, with the Punjab Growth Strategy 2023 setting out plans to focus the curricula more strongly on technical skills and knowledge and improve teacher quality (Planning and Development Board 2019). Punjab could tap on best practices in the Republic of Korea to strengthen the responsiveness of primary and secondary curricula and teaching pedagogy to changing skill needs created by 4IR, as it seeks to strengthen its primary and secondary school education systems (Box 9). In Punjab, the School Education Department is tasked with formulating education policy and maintaining the standards of primary and secondary level education, through formulating the curricula and overseeing the professional development of teaching staff. It can lead efforts to update education curricula and ensure that teachers are provided with sufficient training to deliver the curricula, working closely with the Punjab Planning and Development Board to understand future skill needs.

Box 9: Updating Curricula and Pedagogy to Meet New Skill Needs in the Republic of Korea

The Republic of Korea has one of the most tech-savvy student populations globally. This is reflected by its second-place ranking in the Organisation for Economic Co-operation and Development's (OECD) Program for International Student Assessment (PISA) assessment in 2018 for "digital competence," which is defined by the ability to read, navigate, and understand online texts (OECD 2011).

A key driver behind this is the regular review of the national curriculum framework by the country's Ministry of Education every 5–10 years (National Center on Education and the Economy 2020). This policy seeks to ensure that educational curricula reflect updated learning needs in tandem with the emerging demands of the labor market including new technologies. A revision made in 2015 added six general competencies to students' learning outcomes, one of which is "21st century skills." This includes the introduction of both "soft" skills (such as creative thinking) as well as technical information and communication technology (ICT) skills that the education ministry deems would become increasingly important in the workforce (Kim et al. 2017). The 2015 revision, which had the target of being fully implemented by 2021, also led to the introduction of "creative experiential learning" activities. These are hands-on activities that encourage creative thinking, require mandatory coding classes for all elementary and middle school students, and include ICT training for teachers (Kim et al. 2017).

Apart from ensuring that curriculum is updated, strong emphasis is also placed on ensuring that teachers are updated on pedagogy. The national Professional Development Master Plan, developed in 2015, lays out a comprehensive structure for professional learning throughout a teacher's teaching career in the Republic of Korea. It recommends specific professional learning for educators according to their stage of career development and supports participation in training programs.

Sources: OECD. 2011. Education: Korea tops new OECD PISA survey of digital literacy. https://www.oecd.org/education/educationkoreatopsnewoecdpisasurveyofdigitalliteracy.htm. National Center on Education and the Economy. 2020. South Korea: Learning Systems. http://ncee.org/what-we-do/center-on-international-education-benchmarking/top-performing-countries/south-korea-overview/south-korea-instructional-systems/; Kim et al. 2017. Korea's Software Education Initiative. https://dl.acm.org/doi/abs/10.1145/3151759.3151800?download=true.

C. Industry-Specific Priorities

While the seven recommendations apply to both the textile and garment manufacturing and IT–BPO industries, a set of priorities unique to each industry should be considered when implementing the respective recommendations.

Textile and Garment Manufacturing Industry

Compared to the IT–BPO industry, the textile and garment manufacturing industry has a more limited understanding of 4IR technologies and workers are likely to be lower-skilled and face more challenges in being retrained to take on technical roles (Table 9). The development of an industry transformation road map that addresses the limited awareness of 4IR technologies among firms and sets out measures to increase awareness and deployment must be a priority for policy makers. Efforts to develop an industry-specific road map could be overseen by the Punjab Planning and Development Board, in close collaboration with industry associations such as the All Pakistan Textile Mills Association. Measures to increase the use of 4IR technologies would need to be complemented by policies to ensure that low-skilled workers and female workers can also benefit from the gains of 4IR. Workers in the textile and garment manufacturing would particularly benefit from a robust RPL system that would allow them to seek reskilling and find new jobs in technical fields (e.g., upskill from weaver to production line operator). Female workers could benefit from programs similar to Cambodia's Factory Literacy Program that supports lifelong learning for female garment factory workers (*The Phnom Penh Post* 2021).

Table 9: Summary of Findings in the Textile and Garment Manufacturing Industry

Key Findings
Potential job displacement (% of current workforce): 1.13 million (31%)
Potential job gains (% of current workforce): 1.52 million (41%)
Net job gains from 4IR (% of current workforce): 390,000 (11%)
Top three in-demand skills in 2025: Digital and information and communication technology skills, critical thinking, adaptive learning

Key Challenges	Findings	Recommendations
Limited understanding of 4IR technologies and their applications	Only **35%** of textile and garment manufacturers have a good understanding of 4IR technologies and applications	Develop 4IR adoption road maps for key industries
Workers in manual roles are expected to face a higher risk of displacement while job gains are expected in technical roles	Over **70%** employers expect the proportion of technical roles to increase by 2025	Strengthen existing frameworks for recognition of prior learning
Job gains from 4IR will largely benefit male workers due to low female workforce participation	Only **one-third** of garment workers in Pakistan are female	Adopt inclusive 4IR reskilling policies

4IR = Fourth Industrial Revolution.
Source: Asian Development Bank Sustainable Development and Climate Change Department.

Information Technology–Business Process Outsourcing Industry

The IT–BPO industry has a stronger understanding of 4IR technologies compared to the textile and garment manufacturing industry and skill needs are likely to change rapidly (Table 10). Creative thinking and complex problem-solving skills are expected to become more important to employers in the IT–BPO industry by 2025 and school curricula would need to evolve rapidly to meet these new skills needs. At the same time, it would also be critical for employers and training institutions to work closely together to strengthen the relevance of industry apprenticeships and internships and address the immediate shortage of skilled talent trained in the latest technologies reported by local stakeholders such as the Pakistan Software Houses Association for IT and ITeS (P@SHA). One challenge particularly relevant to the IT–BPO industry is the need to increase the participation of women in STEM careers. This research estimates that of the overall job gains from the adoption of 4IR technologies, the number of jobs gained by male workers will exceed the number of jobs gained by female workers by 9.6 times. In Indonesia, the Philippines, and Thailand, ILO's Women in STEM Workforce Readiness and Development Program aims to enhance the employability of women for STEM-related jobs. The Punjab Planning and Development Board could lead similar initiatives in collaboration with international partners, such as ILO or ADB.

Table 10: Summary of Findings in the Information Technology–Business Process Outsourcing Industry

Key Findings
Potential job displacement (% of current workforce): 15,200 (39%)
Potential job gains (% of current workforce): 22,000 (57%)
Net job gains from 4IR (% of current workforce): 6,800 (18%)
Top three in-demand skills in 2025: Creative thinking and/or design, complex problem solving, adaptive learning

Key Challenges	Findings	Recommendations
Workers would require a significant step-up in creative thinking and design skills	Creative thinking and complex problem-solving skills will become more important to IT–BPO employers	Ensure responsiveness of education systems to changing skills needs
Misalignment on skills demand with training institutions	**75%** of employers plan to adopt augmented reality/ virtual reality (AR/VR) technologies but only **13%** of training institutions offer courses in AR/VR technologies	Strengthen relevance of industry apprenticeships and internships
4IR could exacerbate the challenges faced by female workers in the IT–BPO industry as the number of technical jobs increases	Engineering and other STEM-related careers are associated with field work and long hours in Pakistan and seen as unsuitable for female workers	Adopt inclusive 4IR reskilling policies

4IR = Fourth Industrial Revolution, IT–BPO = information technology–business process outsourcing.
Source: Asian Development Bank Sustainable Development and Climate Change Department.

APPENDIX
Participants in the National Consultations

Table A1: Stakeholders Engaged in Initial Consultations for Punjab, Pakistan

No.	Name	Designation	Organization
Federal Government Ministries and Agencies			
1	Shabnum Sarfraz	Member, Social Sector and Devolution	Planning Commission, Ministry of Planning, Development and Special Initiatives
2	Waheed Zaman	Chief, Mass Media, Culture, Sports, Tourism and Youth Section	Ministry of Planning, Development and Special Initiatives
3	Muqeem-ul-Islam	Director General	National Vocational and Technical Training Commission
4	Aisha Qazi	Director (International Cooperation)	National Vocational and Technical Training Commission
5	Sarwat Ismail	Deputy Technological Advisor	Ministry of Science and Technology
6	Shafqat Abbas	Section Officer	Ministry of Industries and Production
7	Shifa Fatimah	Gender and Child Protection Analyst	Planning Commission, Ministry of Planning, Development and Special Initiatives
8	Tauseef Ahmad	Officer	Planning Commission, Ministry of Planning, Development and Special Initiatives
Punjab Agencies			
9	Ali Salman Siddique	Chair	Technical Education and Vocational Training Authority, Government of Punjab
10	Akhtar Abbas	Director General (Operations-I)	Technical Education and Vocational Training Authority, Government of Punjab
11	Aamer Aziz	Director General (Procurement) / Head of Placement	Technical Education and Vocational Training Authority, Government of Punjab
12	M. Haroon Naseer	Additional Project Director, Skills Sector Coordination Cell, Project Implementation Unit, Punjab Skills Development Project	Industries, Commerce, Investment and Skills Development Department, Government of Punjab

continued on next page

Table A1 *continued*

No.	Name	Designation	Organization
Nongovernment Stakeholders			
1	Rehmatullah Javed	Chair, SME committee	Lahore Chamber of Commerce and Industry
2	Shahid A. Khan	Federation of Pakistan Chambers of Commerce and Industry representative (Chartered Accountant and Director on the Board of Public and Private Ltd. Companies)	Federation of Pakistan Chambers of Commerce and Industry
3	Nabeel-ur-Rehman	Director	Skills Development Council Lahore
4	Ali Faraz Siddiqui	Chief Executive Officer	Aman Institute for Vocational Training
5	Naveed Iftikhar Cheema	Founder	Atoms
6	Gauher Aftab	Chief Executive Officer	Generation Pakistan
7	Sadia Hameed	Senior Project Officer	International Labour Organization

Source: Asian Development Bank Sustainable Development and Climate Change Department.

Table A2: Stakeholders Engaged in Further Consultations for Punjab, Pakistan

No.	Name	Designation	Organization
Federal Government Ministries and Agencies			
1	Ghazala Abid	Chief Industries and Commerce	Ministry of Planning, Development and Special Initiatives
2	Gohar Rehman	Assistant Chief	
3	Aisha Qazi	Director (International Cooperation)	National Vocational and Technical Training Commission
4	Khuram Ahsan	Deputy Director (Planning)	
5	Syed Hassan Mehmood	Joint Secretary	Ministry of Industries and Production
6	Shafqat Abbas	Deputy Secretary	
7	Aisha Humera	Senior Joint Secretary	Ministry of Commerce (Textile Division)
8	Mudassir Raza Siddiqui	Director (Textile)	
Punjab Agencies			
9	Asadullah Faiz	Member, Private Sector Development	Punjab Planning and Development Board
10	Majid Iqbal	Chief (Industries)	
11	Ahmad Saeed	Director General (Finance)	Technical Education and Vocational Training Authority, Government of Punjab

continued on next page

Table A2 *continued*

No.	Name	Designation	Organization
12	M. Haroon Naseer	Additional Project Director, Skills Sector Coordination Cell, Project Implementation Unit, Punjab Skills Development Project	Industries, Commerce, Investment and Skills Development Department, Government of Punjab
13	Sajid Latif	Director General, E-Governance	Punjab Information Technology Board
Industry Associations			
14	Asad Abbas Shah	Research Analyst	All Pakistan Textile Manufacturers Association
15	Saad Umar	Senior Executive Officer	
16	Mujeeb Zahur	Senior Director of Operations at S&P Global Islamabad	Pakistan Software Houses Association (P@SHA)
17	Shafqaat Shah	Senior Director and Head of Commercial Operations at S&P Global Islamabad	

Source: Asian Development Bank Sustainable Development and Climate Change Department.

References

ACT/EMP and ILO. 2017. ASEAN in Transformation: How Technology is Changing Jobs and Enterprises. *Cambodia Country Brief.* https://www.ilo.org/actemp/publications/WCMS_579672/lang--en/index.htm.

Asian Development Bank (ADB). 2020. *Pakistan: Reviving Growth through Competitiveness.* Manila. https://www.adb.org/publications/pakistan-reviving-growth-through-competitiveness.

————. 2021a. *Asian Economic Integration Report 2021.* Manila. https://www.adb.org/sites/default/files/publication/674421/asian-economic-integration-report-2021.pdf.

————. 2021b. *Reaping the Benefits of Industry 4.0 Through Skills Development in Cambodia.* Manila. https://www.adb.org/publications/benefits-industry-skills-development-cambodia.

Aim2Flourish. 2021. The True Definition of Wonder Woman. https://aim2flourish.com/innovations/the-true-definition-of-wonder-woman.

AlphaBeta. 2017. *The Automation Advantage.* https://alphabeta.com/wp-content/uploads/2017/08/The-Automation-Advantage.pdf.

ARY News. 2020. Underprivileged Women to get Smartphones under Kafalat Programme. 31 January. https://arynews.tv/en/poor-women-to-get-smartphones-ehsaas-kafalat-programme/.

Asia Philanthropy Circle and AlphaBeta. 2017. *Catalysing Productive Livelihood: A Guide to Education Interventions with an Accelerated Path to Scale and Impact.* http://www.edumap-indonesia.asiaphilanthropycircle.org/wp-content/uploads/2017/11/APC-Giving-Guide-Book-Final-Report-17112017.pdf.

AT Kearney. 2021. Kearney Global Services Location Index. https://www.kearney.com/digital/article/?/a/the-2021-kearney-global-services-location-index.

AVOXI. 2021. Pakistan Call Center Software Solutions. https://www.avoxi.com/cloud-contact-center-software/pakistan-call-center-solutions/

Baloch, A. 2018. Pakistan, the Next Outsourcing Hub? *Pakistan Today.* 25 June. https://profit.pakistantoday.com.pk/2018/06/25/pakistan-the-next-outsourcing-hub/.

Barber. 2007. Instruction to Deliver: Fighting to Transform Britain's Public Services.

British Council. *Understanding Female Participation in STEM Subjects in Pakistan.* https://www.britishcouncil.pk/sites/default/files/stem_final_with_foreword_1_1.pdf.

Del Buono, A. 2018. Combining Smart Technologies for High Quality. *Control Global.* 29 October. https://www.controlglobal.com/articles/2018/combining-smart-technologies-for-high-quality/.

CB Insights. 2022. The Future of Fashion: from Design to Merchandising, How Tech is Reshaping the Industry. 11 May. https://www.cbinsights.com/research/fashion-tech-future-trends/.

Clarion Technology. How IoT Transforms the Way to a More Sustainable Textile Manufacturing. https://www.clariontech.com/blog/how-iot-transforms-the-way-to-a-more-sustainable-textile-manufacturing.

Clo3d. Design Smarter. https://www.clo3d.com/.

Daily Pakistan. 2020. Punjab Unveils Rs2.22 Trillion Budget. 15 June. https://en.dailypakistan.com.pk/15-Jun-2020/punjab-unveils-rs2-22-trillion-budget.

Daily Times. 2021. Textile Industry of Pakistan in the Time of Covid-19. 16 March. https://go.gale.com/ps/i.do?p=HRCA&u=googlescholar&id=GALE|A655049241&v=2.1&it=r&sid=sitemap&asid=f6fc858d.

———. 2021a. 10 Million Jobs Promise But Not a Single Post on Job Portal. 22 May. https://dailytimes.com.pk/759573/10-million-jobs-promise-but-not-a-single-post-on-job-portal/.

———. 2021b. Parwaaz Launches Four New Skilling Programs to Up Skill National Workforce. 16 June. https://dailytimes.com.pk/773789/parwaaz-launches-four-new-skilling-programs-to-up-skill-national-workforce/.

Dawn News. 2021. Punjab Passes Apprenticeship Act. 19 June. https://www.dawn.com/news/1630123.

Deloitte. 2019. *Empowering Women and Girls in India for the Fourth Industrial Revolution.* https://www2.deloitte.com/content/dam/Deloitte/my/Documents/risk/my-risk-sdg5-empowering-women-and-girls-in-india-for-the-fourth-industrial-revolution.pdf.

Department of Skills Department. Malaysian Skill Certificate (SKM). https://www.dsd.gov.my/index.php/en/service/malaysian-skills-certificate.

Digiskills. Who Should Join. https://digiskills.pk/WhoShouldJoin.aspx.

e-Rozgaar. About Us. https://www.erozgaar.pitb.gov.pk/#erb04.

Economic Times. 2020. AI-Powered Virtual Job Fairs Aim to Create Hassle Free Experience for Job Seekers and Employers. 15 September. https://hr.economictimes.indiatimes.com/news/workplace-4-0/recruitment/ai-powered-virtual-job-fairs-aim-to-create-hassle-free-experience-for-job-seekers-and-employers/78122272.

EdTechReview. 2017. 5 Schools That are Making the Most of AR & VR. 19 September. https://edtechreview.in/trends-insights/trends/2936-5-schools-that-are-making-the-most-of-ar-vr.

European Training Foundation (ETF). 2015. *Promoting Quality Assurance in Vocational Education and Training.* Turin. https://www.etf.europa.eu/sites/default/files/m/B77049AC22B5B2E9C125820B006AF647_Promoting%20QA%20in%20VET.pdf.

European Training Institute. 2021. *Big Data for Labour Market Intelligence.* https://www.etf.europa.eu/sites/default/files/2021-06/guide_en_big_data_lmi_etf.pdf.

Garage, R. 2019. Use Cases of Augmented Reality in Education and Training. 18 January. https://rubygarage.org/blog/augmented-reality-in-education-and-training.

Geo News. 2021. 47% Growth Seen in Pakistan's IT Exports from July–May in Fiscal Year 2020–21. 26 June. https://www.geo.tv/latest/357011-47-growth-seen-in-pakistans-it-exports-from-july-may-in-fiscal-year-2020-21.

Google Cloud Blog. 2021. How to Lower Costs and Improve Innovation with Cloud Computing. 27 July. https://cloud.google.com/blog/topics/research/how-to-lower-costs-and-improve-innovation-with-cloud-computing.

Government of Pakistan, Commerce Division. 2019. *E-Commerce Policy of Pakistan.* Islamabad. https://www.commerce.gov.pk/wp-content/uploads/2019/11/e-Commerce_Policy_of_Pakistan_Web.pdf.

Government of Pakistan, Finance Division. 2021. *Pakistan Economic Survey 2020–21.* Islamabad. https://www.finance.gov.pk/survey_2021.html.

Government of Pakistan, Ministry of Federal Education and Professional Training. 2018a. *National Education Policy Framework 2018.* Islamabad. http://www.mofept.gov.pk/SiteImage/Policy/National%20Eductaion%20Policy%20Framework%202018%20Final.pdf.

———. 2018b. *National "Skills for All" Strategy.* Islamabad. https://www.giz.de/de/downloads/giz2019-en-tvet-national-skills.pdf.

———. 2018c. *Skills for Growth and Development: A Technical and Vocational Education and Training Policy for Pakistan.* Islamabad. https://navttc.gov.pk/national-policies/.

Government of Pakistan, Ministry of Information Technology and Telecommunication. 2018. *Digital Pakistan Policy.* Islamabad.

Government of Pakistan, Ministry of Planning, Development and Reform. 2014. *Pakistan Vision 2025.* Islamabad. https://www.pc.gov.pk/uploads/vision2025/Pakistan-Vision-2025.pdf.

Government of Punjab, Trade Testing Board. Recognition of Prior Learning. https://ttbp.edu.pk/recognition-of-prior-learning-rpl/.

Government of Punjab, Planning and Development Board. 2015. *Punjab Skills Development Sector Plan 2018.* Lahore. https://pnd.punjab.gov.pk/system/files/Punjab_Skills_Sector_Plan_2018_0.pdf.

———. 2019. *Punjab Growth Strategy 2023.* Lahore. https://pnd.punjab.gov.pk/system/files/PGS_2023%2019-21-145.pdf#overlay-context=economic_growth_strategy.

———. 2020. RISE *Punjab: Responsive Investment for Social Protection and Economic Stimulus.* Lahore. https://pnd.punjab.gov.pk/system/files/COVID-19%20Report%289jul2020%29_0.pdf. Government of Punjab, Punjab Information Technology Board. 2018. *Punjab IT Policy 2018.* Lahore. https://policy.pitb.gov.pk/system/files/Punjab_IT_Policy_2018_05062018.pdf.

Graham, P. 2021. Don't Stress About a Job Interview with Bodyswaps' New Simulator. *VR Focus.* 27 May. https://www.vrfocus.com/2021/05/dont-stress-about-a-job-interview-with-bodyswaps-new-simulator/.

Hanaphy, P. 2020. Adidas Reveals Futurecraft Strung, the "Ultimate" 3 Printed Running Shoe. *3D Printing Industry.* 9 October. https://3dprintingindustry.com/news/adidas-reveals-futurecraft-strung-the-ultimate-3d-printed-running-shoe-177073/.

Higher Education Digest. 2020. OnTime Job culminates India's Most Successful Live Recruitment Drive. 11 September. https://www.highereducationdigest.com/ontime-job-culminates-indias-most-successful-live-recruitment-drive/.

HR Technologist. 2019. How AR and VR are Revolutionizing Soft Skills Training in 2019. 22 April. https://www.hrtechnologist.com/articles/learning-development/how-ar-and-vr-are-revolutionizing-soft-skills-training-in-2019/.

International Labour Organization (ILO). 2016. *Wages and Productivity in the Garment Sector in Asia and the Pacific and the Arab States.* Geneva. https://www.ilo.org/wcmsp5/groups/public/---asia/---ro-bangkok/documents/publication/wcms_534289.pdf.

———. 2017. *Employment and Wages Rising in Pakistan's Garment Sector.* Geneva. https://www.ilo.org/wcmsp5/groups/public/---ed_protect/---protrav/---travail/documents/publication/wcms_544182.pdf.

———. 2019a. *Promoting Decent Work in Garment Sector Global Supply Chains.* Geneva. https://www.ilo.org/wcmsp5/groups/public/---ed_protect/---protrav/---travail/documents/projectdocumentation/wcms_681644.pdf.

———. 2019b. *State of Skills: Pakistan.* Geneva. https://www.ilo.org/wcmsp5/groups/public/---ed_emp/---ifp_skills/documents/genericdocument/wcms_742212.pdf.

———. 2020a. *ILO Toolkit for Quality Apprenticeships.* Geneva. https://www.ilo.org/wcmsp5/groups/public/---ed_emp/---ifp_skills/documents/publication/wcms_751114.pdf.

———. 2020b. *What Next for Asian Garment Production after COVID-19?* Geneva. https://www.ilo.org/wcmsp5/groups/public/---asia/---ro-bangkok/---sro-bangkok/documents/publication/wcms_755630.pdf.

InfoTech. 2021. About Us. https://www.infotechgroup.com/about/.

Institute for Workers and Trade Unions. 2020. *Automation and Its Impact on Employment in the Garment Sector of Vietnam.* http://library.fes.de/pdf-files/bueros/vietnam/17331.pdf.

Intellicon. 2017. An Intelligent Call Center Solution from a Pakistani Software Company. 8 September. https://www.intellicon.io/call-center-solution-2/.

Invest Pakistan. Income Tax Exemption on Exports of Computer Software or IT Services or ITes. https://invest.gov.pk/node/1253.

ITE Singapore. Industry Training Schemes. https://www.ite.edu.sg/employers/industry-training-schemes/certified-on-the-job-training-centre.

iWeb. 2020. Back to School: Robotics, Coding Curriculum Pushed Back. 15 January. https://www.itweb.co.za/content/KzQenMj8jrzvZd2r.

Khan, M.Z. 2021. New Textile Policy Envisions Trillion-Rupee Subsidies for Exporters Till Year 2025. *Dawn.* 7 January. https://www.dawn.com/news/1600133.

Kim et al. 2017. Korea's Software Education Initiative. https://dl.acm.org/doi/abs/10.1145/3151759.3151800?download=true.

Lake, K. 2018. Stitch Fix's CEO on Selling Personal Style to the Mass Market. *Harvard Business Review. May.* https://hbr.org/2018/05/stitch-fixs-ceo-on-selling-personal-style-to-the-mass-market.

Lee Kuan Yew School of Public Policy and Microsoft. 2016. *Technical and Vocational Education and Training in Indonesia: Challenges and Opportunities for the Future.* https://lkyspp.nus.edu.sg/docs/default-source/case-studies/lkysppms_case_study__technical_and_vocational_education_and_training_in_indonesia.pdf?sfvrsn=e5c5960b_2.

Malaysian Qualifications Agency. 2019. *Malaysian Qualifications Framework (MQF).* 2nd ed. https://www.mqa.gov.my/pv4/document/mqf/2019/Oct/updated%20MQF%20Ed%202%2024102019.pdf.

Malik, Q.S. 2020. TEVTA Launches Sector Skills Council for Textile. *Daily Times.* 26 December. https://dailytimes.com.pk/706036/tevta-launches-sector-skills-council-for-textile/.

McKinsey & Company. 2012. *Delivery 2.0: The New Challenge for Governments.* https://www.mckinsey.com/industries/public-sector/our-insights/delivery-20-the-new-challenge-for-governments.

———. 2018. *Is Apparel Manufacturing Coming Home?* https://www.mckinsey.com/~/media/mckinsey/industries/retail/our%20insights/is%20apparel%20manufacturing%20coming%20home/is-apparel-manufacturing-coming-home_vf.ashx.

McKinsey Digital. 2020. Reducing Data Costs without Jeopardizing Growth. 31 July. https://www.mckinsey.com/business-functions/mckinsey-digital/our-insights/reducing-data-costs-without-jeopardizing-growth.

Microsoft and AlphaBeta. 2019. *Preparing for AI: The Implications of Artificial Intelligence for Jobs and Skills in Asian Economies.* https://news.microsoft.com/apac/2019/08/26/preparing-for-ai-the-implications-of-artificial-intelligence-for-jobs-and-skills-in-asian-economies/.

Ministry of Trade and Industry. Aerospace. https://www.mti.gov.sg/ITMs/Manufacturing/Aerospace.

———. ITMs Overview. https://www.mti.gov.sg/ITMs/Overview.

Ministry of Information Technology and Telecommunication (MOITT). 2021. *National Freelancing Facilitation Policy 2021.* Islamabad. https://moitt.gov.pk/SiteImage/Misc/files/National%20Freelancing%20Facilitation%20Policy%202021%20-%20Consultation%20Draft%202_0.pdf.

Mustafa, K. 2020. Textile Policy 2020–25: Pakistan to Increase Textile Exports to $25.3 bn by 2025. *The News.* 24 January. https://www.thenews.com.pk/print/603201-textile-policy-2020-25-pakistan-to-increase-textile-exports-to-25-3-bn-by-2025.

Nestlé. 2020. Nestlé Speeds up Factory Support with Augmented Reality. 24 July. https://www.nestle.com/randd/news/allnews/nestle-speeds-factory-support-augmented-reality.

Nicholls-Lee , D. 2021. The Dutch AI Startup Making Online Clothes Shopping More Inclusive. *Dutch News. nl.* 6 January https://www.dutchnews.nl/features/2021/01/the-dutch-ai-startup-making-online-clothes-shopping-more-inclusive/.

No Jitter. 2019. AT&T: Intelligent Pairing Ups Agent Success. 30 January. https://www.nojitter.com/contact-center-customer-experience/att-intelligent-pairing-ups-agent-success.

Olson, P. 2018. Google, Microsoft and Startups are Going to War on Chatbot Technology. *Forbes.* 27 July. https://www.forbes.com/sites/parmyolson/2018/07/27/google-microsoft-and-startups-are-going-to-war-on-chatbot-technology/?sh=4ed1840461b6.

Organisation for Economic Co-operation and Development (OECD). 2012. *Skills Development Pathways in Asia.* https://www.oecd.org/cfe/leed/Skills%20Development%20Pathways%20in%20Asia_FINAL%20VERSION.pdf.

———. 2011. Education: Korea tops new OECD PISA survey of digital literacy. https://www.oecd.org/education/educationkoreatopsnewoecdpisasurveyofdigitalliteracy.htm.

P@SHA. Inauguration of the P@SHA Skill Development Program 2021–2022. https://www.pasha.org.pk/pasha-skill-development-program-2021-22/.

Pakistan Today. 2020. Imran Launches Rs30bn Hunarmand Jawan Program. 9 January. https://profit.pakistantoday.com.pk/2020/01/09/pm-imran-launches-rs30bn-hunarmand-jawan-programme/.

Pakistan Bureau of Statistics. Labour Force Statistics. https://www.pbs.gov.pk/content/labour-force-statistics.

———. Industry. https://www.pbs.gov.pk/content/industry.

Pro Pakistani. 2015. InfoTech Helps Ghana Automate Its Stock Exchange. https://propakistani.pk/2015/11/10/infotech-helps-ghana-automate-its-stock-exchange/?__cf_chl_jschl_tk__=pmd_6WRsbcsedyiuPl29E1zbgPmEC5Hdih7SI005pRXzQ9w-1629426914-0-gqNtZGzNAiWjcnBszQml.

PSDF. 2020. PSDF and Coursera Launch Free International Online Learning Courses. 9 November. https://www.psdf.org.pk/psdf-and-coursera-launch-free-international-online-learning-courses/.

Punjab Board of Investment and Trade. Textile Industry. http://www.pbit.gop.pk/textile_ind.

Punjab Skills Development Fund. https://www.psdf.org.pk/.

———. Training Service Providers. https://www.psdf.org.pk/tsp/industry/.

Radio Pakistan. 2020. TEVTA starts e-learning courses in wake of COVID-19: Chairman. 28 July. https://www.radio.gov.pk/28-07-2020/tevta-starts-e-learning-courses-in-wake-of-covid-19-chairman.

Qasin, M. 2021. $150 Million Revenue Brought by "Pakistani Freelancers" in a Year. 15 March. https://startuppakistan.com.pk/150-million-revenue-brought-by-pakistani-freelancers-in-a-year/.

Reddy, T. 2017. How Chatbots Can Help Reduce Customer Service Costs by 30%. *IBM Watson Blog*. 17 October. https://www.ibm.com/blogs/watson/2017/10/how-chatbots-reduce-customer-service-costs-by-30-percent/.

Saeb, P. 2017. Virtual Reality Potential for Training Contact Centre Agents. *Contact-centres.com*. 3 January. https://contact-centres.com/virtual-reality-potential-training-contact-centre-agents/.

Schmelzer, R. 2019. AI Applications in Education. *Forbes*. 12 July. https://www.forbes.com/sites/cognitiveworld/2019/07/12/ai-applications-in-education/?sh=13bf1b6b62a3.

Shaham, H. 2020. Augmented Reality Customer Service—A Success Story. *TechSee Blog*. 9 March. https://techsee.me/blog/augmented-reality-customer-service/.

Skurio. 2020. Skurio Research Reveals Latest Trends in Cybersecurity Challenges. 20 August. https://skurio.com/press/skurio-research-reveals-latest-trends-in-cybersecurity-challenges/.

Study Solutions.PK. PSDF Free Garments & Fashion Designing Courses With Monthly Stipend 2022. https://www.studysolutions.pk/free-garments-fashion-designing-courses-with-monthly-stipend.

Thakur, N. 2016. Robots vs People? Automation Set to Hit Jobs in India's Textile Industry. *Catch News*. 8 July. http://www.catchnews.com/business-economy-news/robots-vs-people-automation-set-to-hit-jobs-in-india-s-textile-industry-1467996260.html.

The Nation. 2020. TEVTA Implementing Ambitious Programme to Produce Skilled Workers. 27 December. https://nation.com.pk/27-Dec-2020/tevta-implementing-ambitious-programme-to-produce-skilled-workers.

The News. 2019. Pakistan Produces 20,000 IT Graduates, Engineers Annually, Says Minister. 2 November. https://www.thenews.com.pk/print/549580-pakistan-produces-20-000-it-graduates-engineers-annually-says-minister.

———. 2020. Pakistan's IT Exports to Cross $1.2bln This Fiscal Year. 2 December. https://www.thenews.com.pk/print/752228-pakistan-s-it-exports-to-cross-1-2bln-this-fiscal-year.

The News International. 2018. Sector Skills Councils Launched to Create Employment. 24 February. https://www.thenews.com.pk/print/284976-sector-skills-councils-launched-to-create-employment.

The Phnom Penh Post. 2021. Factory Literacy Programme Expands. 29 March. https://www.phnompenhpost.com/national/factory-literacy-programme-expands.

Today. 2018. New Jobs Portal Seeks to Reduce Job–Skills Mismatch. 17 April. https://www.todayonline.com/singapore/new-jobs-portal-seeks-reduce-job-skills-mismatch.

Tribune. 2021. HEC Policy to Overhaul University Education. 15 March. https://tribune.com.pk/story/2289476/hec-policy-to-overhaul-university-education.

Tukatech. 2021. Levi's Largest Pakistan Knit Supplier Expands Capacity with TUKA Cutter. 14 May. https://tukatech.com/combined-fabrics-third-fabric-cutter/.

United Nations Children's Fund (UNICEF). 2019. *Developing Skills in Youth to Succeed in an Evolving South Asian Economy—A Case Study on Pakistan.* New York. https://www.unicef.org/rosa/media/4566/file/Pakistan_Youth_Skills_Case_Study%20.pdf.

United Nations Educational, Scientific and Cultural Organization (UNESCO). *Education 2030: Incheon Declaration and Framework for Action for the implementation of Sustainable Development Goal 4.* Paris. http://uis.unesco.org/sites/default/files/documents/education-2030-incheon-framework-for-action-implementation-of-sdg4-2016-en_2.pdf.

Virtusize. 2021. About Us. http://w.virtusize.jp/site/.

Waya. 2019. How IOT Is Transforming the Garment Industry. 10 April. https://waya.media/how-iot-is-transforming-the-garment-industry/.

WNS. Does the Cloud Come with a Silver Lining for BPO? https://www.wns.com/insights/articles/articledetail/81/does-the-cloud-come-with-a-silver-lining-for-bpo.

Workforce Singapore. Career Conversion Programmes (CCP) for Individuals. https://www.wsg.gov.sg/programmes-and-initiatives/career-conversion-programmes-individuals.html.

World Economic Forum. What is the Fourth Industrial Revolution? https://www.weforum.org/agenda/2016/01/what-is-the-fourth-industrial-revolution/.

Xinhua. 2021. Feature: Pandemic-Promoted Online Shopping Becoming New Normal in Pakistan. 13 February. http://www.xinhuanet.com/english/asiapacific/2021-02/13/c_139741338.htm.

zSpace. Beyond STEM: Building Soft Skills with Augmented and Virtual Reality. https://zspace.com/blog/going-beyond-stem-building-soft-skills-with-augmented-and-virtual-reality.

www.ingramcontent.com/pod-product-compliance
Lightning Source LLC
Chambersburg PA
CBHW050047220326
41599CB00045B/7319